INNOVATE ENERGY

THE MICROREACTOR REVOLUTION— TRUTH, GRIT, AND THE FUTURE OF AMERICAN POWER

INNOVATE ENERGY

THE MICROREACTOR REVOLUTION—
TRUTH, GRIT, AND THE
FUTURE OF AMERICAN POWER

MIKE WANDLER

abundance
collective

DEDICATION

To Jerry Schloredt,
my father-in-law,
a quiet warrior of the atom.

You ran nuclear microreactors—not for fame or fortune,
but for country, for progress, and for the future—
in the heat of Panama and the ice of Antarctica,
under the Seabee banner, where few dared to go.

You lived the principles of resilience, precision, and integrity,
long before they were buzzwords in energy innovation.
This book carries your spirit forward—
from field-tested microgrids to the bold dreams
of powering industries and tribes in flow.

Thank you for showing me what it means
to serve with grit, to lead with heart,
and to light the way for others
even in the coldest, darkest places.

This is for you, Jerry.
The mighty micro reactor will finally hum again.

TABLE OF CONTENTS

PART I—THE ENERGY CRISIS & THE ENTREPRENEURIAL RESPONSE

PART II—INNOVATION, LEADERSHIP, AND THE NEW ENERGY PLAYBOOK

PART III—THE HUMAN FACTOR

PART IV—THE MANIFESTO

FOREWORD

Rarely does history come full circle.

In 1960, I began my Navy Nuclear Shore Power career at the Advanced Nuclear Power School in New London, CT, then went on to train and qualify operationally on two microreactors in the Army Nuclear Power Program (ANPP). The first was a stationary plant, the SM-1, at Ft. Belvoir, VA, and the second, the PM-2A, operated by the Army at Camp Century, Greenland, in tunnels cut into the ice cap.

In 1963, as a young LT Fegley, I was given the privilege, and the challenge, along with a mature, highly trained enlisted crew, of becoming the Officer in Charge of the Navy's PM-3A Nuclear Power Plant at McMurdo Station, Antarctica, providing power for the National Science Foundation, Operation Deepfreeze. This microreactor had been designed, constructed, transported to Antarctica, erected, and was taken critical for the first time in little more than two years. Just six months later, it was powering the Station.

For the first two Antarctic winters, the PM-3A operated under the responsibility of a contractor with a highly trained and specialized crew of Seabees, Hospital Corpsmen, and four men from the Army program. Our crew, Crew III, was highly qualified and experienced for our mission. We had the good fortune to be the crew to assume responsibility for the plant from the contractor for the Navy after the final test run, which was completed on the last day before the beginning of the Austral winter, March 5, 1964. We weren't there for fame or headlines. We were there to serve our country, support the scientists, produce power, and prove that safe, reliable nuclear power could be delivered anywhere on earth—even at the bottom of the world.

Our team faced the harshest conditions imaginable. And, we did it! The PM-3A microreactor delivered clean power through some of the worst that Antarctica could throw at us: six months of darkness, minus-60 degrees plus wind chill, and zero room for error. Every day was a test of ingenuity, discipline, and teamwork. We were there to prove to our Navy superiors that with our education, training, experience, and special skills, a nuclear reactor could be operated safely, remotely, and without direct oversight. Our mission was clear: keep the reactor operating safely, the lights on, and make sure every man came home safely. We proved that microreactors weren't just possible—they were a safe, reliable energy source. It was my team, some of the best in the Program; dedicated men who met every challenge with the Seabee "Can Do" attitude that made success a daily reality.

The commemorative plaque marking the site of the PM-3A nuclear power plant (United States Antarctic Program)

After fourteen months deployed at McMurdo Station, I returned to the Naval Nuclear Power Unit at Fort Belvoir, VA, to support the PM-3A for another year and a half. Several years later, I returned to the microreactor program and served for another three years as the Officer in Charge of the Naval Nuclear Power Unit, responsible for the PM-3A and numerous radioisotope power generators.

It's a point of pride and a frustration that, after achieving so much in microreactor development and demonstrating their potential, politics and short-term thinking led America to abandon microreactors just as we were getting started. The technology worked. The teams performed, but momentum was lost, and for decades, this remarkable solution to clean, reliable power has been left on the shelf.

That's why this book matters so much.

Continuing, I'd first like to highlight the fact that Mike Wandler's father-in-law, Jerry Schloredt, served as our best PM-3A reactor control room operator, then an Operations Supervisor, and finally Chief Petty Officer in Charge of the last crew. Yes, that's three winters and a summer support tour at the PM-3A, plus a tour on the Army's barge plant in Panama. It was through Jerry, my fishing buddy in later years, that I had the good fortune to meet Mike and tour some of his impressive industrial facilities. What impressed me most was their size, the latest high-tech machines, and his leadership as a manager.

Mike and his team are picking up where we left off—honoring the legacy of our nuclear pioneers and finally carrying this mission forward. This charge couldn't be spearheaded by better hands. For those of us who built and ran those early microreactors, it's a thrill to see the next generation get the chance to do what we always knew was possible: deliver safe, clean, resilient energy wherever America needs it; for industry, for national security, and for the future.

To the builders, engineers, and entrepreneurs reading this: You're not just reviving an old idea. You're building on a foundation tested in the harshest environments by people who believed that "good enough" was never good enough. You are finishing the work we started—taking microreactors out of the history books and putting them back in the field, where they belong.

My advice? Stay true to the mission. Do the designs, select and focus on the best technology, construct the prototypes to prove a concept, and gear up for full production, led by companies like L&H, who can get the job done. Never compromise on safety, quality, or integrity. WE CAN'T AFFORD AN ACCIDENT! Remember, real innovation is never easy, but it's always worth it. The world will throw obstacles in your path. Don't let politics or fear slow you down. Build for the next generation. Lead with courage, heart, and above all, Safety!

To see microreactors humming again—to know that America's "can-do" spirit is alive and well—means more to me than I can say. On behalf of every ANPP Nuke who braved the ice, the heat, and the unknown: We're cheering you on. It's about time.

Lt. Charles E. Fegley III, CEC, USN (Ret.Captain) Officer in Charge, PM-3A Microreactor, McMurdo Station, Antarctica, October 1963-November 1964

A NOTE TO THE READER

Our world is at an energy crossroads. After all the debates about climate, technology, and economics, one thing is clear: **we do not lack resources, intelligence, or opportunity. We lack leadership with the courage to build boldly and honestly, with a long-term vision in mind.**

At L&H, a global heavy industrial manufacturer and repair business, and at Evercore, a leader in energy solutions, we don't believe in chasing the cheapest answer or waiting for permission. My Unique Ability—and the core of our approach—is entrepreneurial leadership. Through that, we aim to be the private partner to government, academia, and industry, delivering solutions that are real and at the speed of business.

For sixty-one years, L&H has built and rebuilt the biggest machines on earth under real deadlines with real financial accountability. The disciplines needed to feed the nuclear supply chain—material pedigree, process control, NDE, repair/refit at speed, and lifetime support—are L&H's lane. We make "Safe. Quality. Competitive." real every day.

Reactors don't run on press releases; they run on certified steel, traceable welds, and maintainable assemblies built by people who live or die by first-pass yield. We have plenty of ideas and capital. We need manufacturing capacity, qualifications, and uptime. That's what gives me an edge in the energy conversation.

I'm not writing to simply make a case for microreactors. I'm putting out a call to rebuild American industrial capability so microreactors actually ship. I want to reveal what happens when you put builders, doers, and problem-solvers in charge of energy's future. This book will demonstrate the power of partnership, learning, and relentless iteration of best practices from those who came before. It's about the relentless pursuit of **energy abundance** for our nation and our allies in the free world.

Why would I publish my playbook? Shouldn't I worry about people stealing my ideas?

A few years ago, I realized my life's purpose is to innovate energy for the good of mankind. My passion is to envision and manifest the future, leading like-minded people with a clear mission, vision, and values. I love being the CEO of this company that my father started. It has been such a fun and rewarding endeavor. L&H uses some very unique, state-of-the-art equipment to make parts for the biggest machines on earth. I started working here over forty years ago. I love my tribes and their tribes—my employees, customers, and vendors—and I plan to do this for another forty years and beyond. I feel like I'm just getting started.

But even with the satisfaction I feel running L&H, I've found sharing and collaborating with like-minded people to innovate energy is my superpower, and there's enough demand for hundreds of companies to share best practices and execute.

My friend Peter Diamandis from Abundance 360 says, "No scarcity, just abundance." I don't need to hoard my ideas. The more I collaborate, the bigger this can be. Another friend and M.I.T.'s Entrepreneurship leader, Bill Aulet, believes "stealthy is unhealthy." Keeping my vision under the cover of darkness until it's fully ready won't let it grow. I invite you to join me as I manifest the innovation of energy for mankind. This book will let you see what it looks like. Entrepreneurship and innovation are messy, so hang on tight; this will be a wild ride.

THE THREE DRIVERS OF ENERGY ABUNDANCE

Energy is a hot topic wherever you go. But I want you to forget the old policy talk and buzzwords. In the real world, three things determine whether an energy solution works for the people who depend on it. I call safe, quality, and competitive the Three Drivers of Energy Abundance. These resonate with me because L&H considers them our top three priorities when it comes to heavy industrial manufacturing and repair. Not only do they stand the test of time, but they also have a place of significance across industries.

SAFE, QUALITY, AND COMPETITIVE

In the energy sector, the triad of the Three Drivers must be addressed. To improve energy conditions, we need a plan that treats every region independently but considers each piece, weighing them according to the relevance in their community.

- **SAFE**: When we think about safety, we need to include the needs of people, the environment, and all living things on the planet. Every energy source carries risks; however, we must compare those risks fairly, side by side, without hype or hidden agendas. We can't ignore, hide, or exaggerate risks for the sake of commercial or political spin. If any energy solution creates extraordinary dangers compared to alternatives that outweigh its benefits, we don't want any part of it. I haven't come across anyone in any energy industry who doesn't care about clean and safe.

- **QUALITY:** "Good enough to get by" doesn't cut it in my book. My teams demand solutions built to last. They have to be reliable and durable, consistently delivering real value. Quality means no surprises, no shortcuts, and no hidden costs down the road. When we discuss quality, we're talking about meeting or exceeding customer expectations for the entire lifecycle.

- **COMPETITIVE:** This does not mean being the cheapest. Cheap energy, like cheap anything, usually ends up costing more in the long run. Remember, if you can't afford to do it right the first time, you sure as hell can't afford to do it twice.

These Three Drivers shape every decision we make at L&H and Evercore, and they're the lens I used when writing this book. Without all three pieces tailored to specific regions, any energy we harness will fall short.

LEGACY & INSPIRATION: THE REAL MICROREACTOR PIONEERS

Most people think microreactors are a new idea. But decades before the buzzword "modular nuclear" became popular, American teams proved it in the field. They went to the edge of the world, far from any city or grid, and lived on zero backup for six months. Nothing on earth compares to the remoteness these pioneers experienced in the 1960s. Their journey compares to putting a nuclear reactor on the moon today.

HISTORICAL PRELUDE: THE UNTOLD STORY OF MICROREACTORS

In the late 1950s and 1960s, the US military quietly launched one of the most daring energy experiments in history: the Army Nuclear Power Program (ANPP). Between 1957 and 1967, it designed, built, and operated eight small reactors in some of the most remote, unforgiving environments on Earth.

- **SM-1 (1957, Fort Belvoir, Virginia):** The prototype, proving that a small nuclear power plant could be built and run safely inside the United States.

- **PM-2A (1961, Camp Century, Greenland):** Buried under the Greenland ice cap, it provided heat and electricity to an Army base carved into tunnels of ice.

- **PM-1 (1962, Sundance, Wyoming):** The first reactor designed to fit into a C-130 cargo plane, it powered a remote Air Force radar station.

- **PM-3A (1962–1972, McMurdo Station, Antarctica):** Perhaps the boldest of all, this plant was designed, shipped, erected, and producing power in little more than two years—keeping the US science base alive through Antarctic winters and even desalinating seawater.

- **MH-1A (1967, Panama Canal Zone):** A floating nuclear plant aboard the barge *Sturgis*, it powered the locks of the Panama Canal.

- **Other prototypes** at Fort Greely, Alaska, and other test sites further refined the technology.

These weren't experiments on paper—they were operational power plants. They delivered safe, resilient energy where no diesel supply line could be guaranteed. Crews endured minus-60°F wind chills in Antarctica, drilled deep into Greenland's ice cap, and maintained floating reactors in tropical Panama. Every deployment stretched logistics, engineering, and human endurance.

The results were undeniable: **microreactors worked.** They provided clean power to bases that would have otherwise been impossible—or prohibitively expensive—to sustain with fuel shipments.

And yet, despite this success, momentum was lost. Political winds shifted. Budgets tightened under the strain of the Vietnam War. The ANPP was quietly shelved, and America walked away from microreactors just as they were proving their worth.

LEADING AT THE EDGE: LT. CHARLES E. FEGLEY III, CEC, USN

Imagine being dropped off at the bottom of the world on a massive sheet of ice, knowing you wouldn't see another ship for six months. If the reactor you were responsible for went down, it was over. Heat, water, food—everything crucial for survival—depends on your crew producing energy.

In 1964, Lt. Charles E. Fegley III of the US Navy Civil Engineer Corps (CEC) was chosen as Officer in Charge of the Navy's PM-3A microreactor at McMurdo Station, Antarctica. As the commanding officer for "Deep Freeze '64" Winterover, Chuck led the Seabee detachment and support personnel who operated and maintained the PM-3A microreactor, delivering continuous heat and safe nuclear power in the most unforgiving environment on earth.

Navy records and news articles cite Chuck as the responsible officer for safe operation and crew leadership. In early 1965, after heading up the reactor's refueling, he handed over command. Chuck Fegley's leadership makes him one of the first American officers ever to command a deployed microreactor team—proving, long before it was trendy, that nuclear power could be portable, safe, and reliable anywhere on earth.

To recognize his contribution, a tributary glacier is named after him. You'll find the Fegley Glacier in the Holland range in Antarctica.

A LEGACY CARRIED FORWARD: SENIOR CHIEF PETTY OFFICER JERRY SCHLOREDT

My father-in-law, Senior Chief Petty Officer Jerry Schloredt of the US Navy Seabees' nuclear division, served with Lt. Fegley, also pushing the limits of portable nuclear in the harshest conditions. With a Senior Reactor Operator certification, Jerry operated microreactors in Antarctica and Panama, supporting military installations and scientific research where nothing else could do the job. His technical excellence, discipline, and commitment to safety kept these reactors running—often under conditions that demanded grit and calm leadership — earned him a place in history as well as a spot on the map. The Schloredt Nunatak, a ridge in the Perry Range of Marie Byrd Land in Antarctica, bears his name.

WYOMING ROOTS AND A FAMILY LEGACY

In addition to the influence of people like Chuck and Jerry, I'm also drawn to this field of nuclear power because of its Wyoming roots. The United States built America's first land-based microreactor, PM-1, in 1962, just outside of Sundance, about an hour from my hometown of Gillette.

And the impact of the Sundance reactor extends beyond purely professional considerations. Jerry moved his family to the Sundance area before he left for Antarctica to work on the PM3-A reactor. That's how I met my lovely wife, Jerri.

Jerry's stories and Chuck Fegley's leadership shaped my view of what's possible with nuclear microreactors. I connected with some of the Navy Nuke team at Idaho National Laboratory a few years ago. They all told me they were blown away when the microreactors were abandoned just as they were ready to make a significant impact in the 70's.

Meeting and listening to the old guard share their stories at the nuclear research facility was a tremendous step in achieving this mission of innovating nuclear energy. Jerry's story and his influence definitely helped me narrow the focus of my purpose.

A Roadmap for the Future

What Chuck and Jerry and their Navy Nukes teams accomplished wasn't just historic; it's a blueprint for what comes next. Decades before anyone else talked about microreactors, they proved that small, resilient, safe nuclear power is possible, and their standards for safety and leadership still set the bar for innovators today.

This is why today's innovators are not inventing something new. They are reviving a proven solution, one that pioneers like Jerry Schloredt, Chuck Fegley, and their crews demonstrated six decades ago. Where politics abandoned the promise, industry now has the chance to finish the job.

As microreactors make their comeback, with new projects at Idaho National Lab like PELE, MARVEL, EVINCI, KALEODIS, it's their example that inspires me—and should inspire everyone—to finish what they started.

Maybe you're wondering why I'm so passionate about this mission. It's because I've seen up close what's possible when leaders like Chuck Fegley and Jerry Schloredt are in charge. This isn't nostalgia. It's a legacy that points the way forward—back to the future, and beyond.

The mission ahead is clear: to take microreactors out of the history books and scale them for industry, security, and communities across the globe. The pioneers showed us what was possible. Now it's time to deliver on that promise.

PART I

THE ENERGY CRISIS &
THE ENTREPRENEURIAL
RESPONSE

CHAPTER 1

THE ENERGY CRUNCH—
AMERICA'S NEXT BOOM,
WHO GETS CUT, AND
WHY MICROREACTORS
ARE THE FUTURE

B ob woke to his cell phone vibrating on the stand next to him. *It must be eighty-five degrees in here.* Checking the air conditioning, he found it was barely pushing cool air. *We're supposed to have record heat this week. Perfect day for the AC to go.* His phone buzzed a second time, but the notification wasn't a text or social media as he expected. Instead, the screen had a national alert.

Please reduce power use. Rolling blackouts likely this afternoon.

THE END OF "INVISIBLE" ENERGY

Bob's story sounds futuristic; however, it could happen tomorrow. You might expect to find this news in California or Texas, but it will hit much closer to home. Wyoming, Ohio, Pennsylvania—even the place you live has been impacted.

But what if you're running more than a household? How will this affect you if you're running a billion-dollar mine, a state-of-the-art chip foundry, a hospital, or a next-generation AI data center?

Let's be clear: When the grid gets tight, it's not your neighbor who gets cut first. Industry, mining, and big energy users always move to the front of the line to have power curtailed. The kids and grandmas get the juice, as they should. Your business gets the call, "Shut down now, or we'll shut you down."

WHEN THE GRID GETS TIGHT, IT'S NOT YOUR NEIGHBOR WHO GETS CUT FIRST.

For a generation, America's energy "miracle" has been invisible. Reliable and affordable, every time you flipped the switch, power appeared. Unless you climb poles or report to the energy production site every day, you never think about it.

That era is over.

For the first time in over forty years, our national energy demand isn't flatlining. It's exploding.

AI chip factories are springing up from Arizona to Wyoming, each one demanding as much electricity as an entire city—or even a whole state.[1] In Cheyenne, Wyoming, a new AI plant is coming online. The promise of one hundred full-time jobs sounds attractive. But there's a catch. The operation will use as much electricity as the entire state, and the grid simply can't supply it. The company is forced to bring in its own power—massive natural gas turbines for now, but tomorrow, who knows? Microreactors may be the solution.

This isn't an isolated project; it's a wave. Intel, TSMC, Samsung, and others are building mega-chip plants from Texas to New York, all demanding gigawatts of new power, enough to crash the local grid if they all plug in at once. And that's before you consider the energy needs of "reshoring" manufacturing,

30

minerals, and mining, or the next round of electric vehicles, data centers, and advanced industrial automation.

DOUBLING AMERICA'S ENERGY SUPPLY ISN'T OPTIONAL

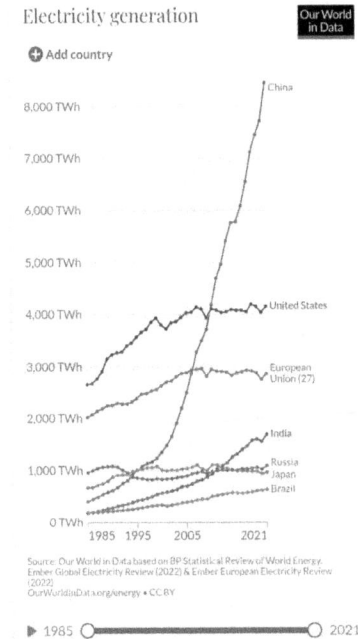

Electricity generation

Let's put the numbers in black and white:

- US electricity production has barely grown in 30 years.
- Meanwhile, China has built an additional 8,000 terawatt-hours of generation since 1990— enough to power the entire United States, with power to spare. (See chart below.)
- According to the best industry estimates, the United States must DOUBLE its electricity production

over the next fifteen to twenty years just to meet the demands of AI, chips, and digital infrastructure—not to mention new factories, critical minerals, or the dream of electrifying everything.

This is not hype. Microsoft, Google, and Amazon all signal the same thing:

- Data is the new oil, and electricity is the new national security.
- If you don't have enough power, you can't win the AI race. You can't win the chip race. You can't keep your industries, let alone your country, competitive.

Here's a hard truth most people never see until it's too late: When power runs short, utilities and grid operators have to make a choice. Will they power citizens, hospitals, or industry? It's no surprise to find that citizens, hospitals, and critical infrastructure come first, while industry and big business get the leftovers—or nothing. When the grid is on the brink, the public must be protected.

But for mines, mills, foundries, chip plants, or major industrial sites, this means they are always shut down first when things get tight. The ramifications of this practical protection of humanity become costly:

- Machines and computers melt down or burn up if the voltage drops.
- A 440-volt machine running at 340 volts doesn't just slow down; it destroys itself.
- When heavy industrial, high-use facilities don't turn off power fast enough, they end up with fried

equipment, lost production, missed deliveries, and wrecked financials.

Some utilities are courteous enough to call you with a warning. But if someone falls asleep at the switch, things move too fast, or the utility company just doesn't care, businesses get hit hard. And downtime for a mine, a foundry, or a chip plant can cost millions per day.

If you own one of these businesses, you can't take the risk. You *must* find your own power source—one that the grid can't curtail, no matter what. This is how we build the future.

ENERGY ON YOUR TERMS

For the first time since the nuclear navy, American business needs energy that is:

- Resilient
- Local
- Always on
- Not at the mercy of grid curtailments or government mandates

Microreactors are the only technology with the power density, safety, and reliability to fill this gap. And they're not just for remote mines or Arctic research stations. They will fill the need for every business, factory, or industrial site that can't afford to be "second priority" anymore.

- Chip plants in Wyoming, powered by their own micro-reactors, will be immune to blackouts, brownouts, and politics.

- New mines and mineral processing plants will go wherever the minerals are, instead of where the grid is strong enough.

- Hospitals, universities, and data centers will never have to worry about the grid because their own microreactors power everything 24/7 for decades at a time.

This is not science fiction. It's the next energy revolution. And it's already begun.

THE CHINA COMPARISON: BUILDING WHILE WE WAITED

While America debated solar vs. wind, China built everything—coal, gas, hydro, nuclear, solar, wind, and transmission lines. They more than doubled their national power supply, creating 8,000 terawatt-hours of new generation in a single generation. Now China leads the world in minerals, manufacturing, chips, and digital infrastructure. Meanwhile, American companies are forced to bring their own power because the current grid simply can't keep up.

LESSONS FROM THE FRONTLINES: WHY THIS BOOK MATTERS

I've spent my career in the real world of energy. Conducting business around the world, I've experienced blackouts that were more than just inconvenient; people lost jobs, fried equipment affected the budget, and livelihoods were on the line. I watched my father build L&H Industrial with his own hands, serving the biggest machines in Wyoming. I listened to stories about how my father-in-law, Jerry Schloredt, ran nuclear microreactors in Antarctica and Panama, with no backup or safety net—just relentless innovation and grit. Today, my team at L&H and Evercore is preparing for the next boom. We work with mining, manufacturing, and AI leaders who know the grid won't protect them. We see the demand coming—the AI plants, chip factories, and critical industries that can't afford to be left in the dark. We know the only way forward is Entrepreneurial Energy: a new era of American leadership that includes microreactors and distributed solutions.

A WARNING AND A CALL TO ACTION

Some people prefer to pretend that wind and solar alone can replace all legacy power. But if we continue to ignore the value of microreactors, we will lose the next industrial revolution. Germany tried it. They shut down nuclear power and doubled down on renewables, driving energy costs so high that their industries nearly collapsed. They're now burning more brown coal than ever and pretending it's still "green." If America will learn from its mistakes, it can save the sunk costs of doing what's popular and letting social media lead energy policy.

The real future is an "all of the above" energy mentality. By putting nuclear, solar, wind, and water on the table alongside legacy energy, we can have smart, competitive, reliable energy that can be dispatched where and when it's needed, with the consumer's ability to pay always front and center.

The mindset shift begins by recognizing microreactors are not a luxury. They are a necessity for mines, mills, chip plants, and industries that power America, as well as for national security. Without bringing nuclear energy into the forefront, we become dependent on Chinese or Russian energy. Nuclear is also necessary for the families, employees, and communities who deserve more than just "good enough."

THIS IS THE NEW ENERGY BATTLEGROUND

This book is your entrepreneurial leadership playbook, written to spread the truth about energy. It's my mission, vision, and values that drive my life's purpose to innovate energy. I write for builders, doers, and leaders who refuse to let America and the free world fall behind. And most importantly, I share for the good of the human race and all living things we participate with in this world.

The Bottom Line:
The energy crunch is here.
The next boom is waiting.
Who gets cut? Who gets to build?
Who gets to lead the future?
It's time to get to work.

CHAPTER 2

THE REAL PRICE OF ENERGY

When people talk about energy policy, they throw around words like "affordable" and "accessible," as if those are the only things that matter. But in my world of heavy equipment, manufacturing, rural towns, and working families, those words don't mean much unless you understand who actually pays the price when things go wrong.

The truth is, the middle class, businesses that can't pass the cost to someone else, and communities left on the edge of the map pay the real cost of energy. Every time a regulator decides to play it safe by adding another layer of bureaucracy or a utility company cuts corners to hit a cost target, it's regular people—not lobbyists or think tanks—who pick up the tab.

CHEAP ISN'T THE SAME AS COMPETITIVE

> THE CHEAPEST FIX USUALLY ENDS UP BEING THE MOST EXPENSIVE ONE.

In the race to lower rates, too many decision-makers chase "cheap" at the expense of "value." That's a mistake. Here's what I learned from my dad at L&H: The cheapest fix usually ends up being the most expensive one. When you buy junk, you pay for it twice—once up front, and again when it fails at the worst possible time.

The same goes for energy. Real competitiveness isn't about being the lowest bidder—it's about providing the best total value over time. That means reliability, uptime, long-term support, and resilience when disaster hits.

We have to focus on more than just rates. Who gets left holding the bag when things break down? Too many policies are written by people who never have to live with the consequences. When the grid goes down, who pays?

- The rancher whose pumps freeze.
- The mining operation, whose million-dollar haul stops.
- The family who watches their food spoil and their pipes burst.

Meanwhile, the people who made the rules or "optimized" the system are long gone, protected by contracts, lawyers, or public funding.

THREE DRIVERS
SAFE
QUALITY
COMPETITIVE

That's why our Three Drivers—Safe, Quality, and Competitive—are more than just business principles. They're a promise to those who can't pass the bill on to anyone else. It's why I don't buy the idea that "affordable" alone is good enough. If it's not safe, if it's not high quality, if it doesn't compete in the real world, then it's not good enough for the people who matter.

THE TRUE COST OF BROWNOUTS AND BLACKOUTS

Earlier, I mentioned the cost of brownouts. But if you want to see the real, hidden cost of bad energy policy, dive deep into what happens when the utility companies turn off the power. You'll discover the consequence is catastrophic loss.

Brownouts and blackouts do more damage, more quickly, than any price hike ever could. Equipment gets destroyed, production lines grind to a halt, hospital patients, miners, and everyday families are put at risk, and jobs vanish overnight.

I've seen it firsthand. When L&H opened a factory in South Africa, we thought we had every angle covered. But then a well-meaning government decision changed everything. They decided to take all the coal off the trains and put it on the roads to increase trucking jobs.

It didn't play out like they hoped.

- The roads couldn't handle the traffic; they were destroyed in months.
- With the coal supply chain crippled, power plants couldn't keep up.
- Rolling brownouts became the new normal, and the grid's reliability collapsed.

The efficiency and affordability of our operations didn't matter. The outage wrecked our machines, and we lost orders. Our people couldn't count on steady paychecks or safe working conditions, and we couldn't quote firm delivery dates. The business became unprofitable overnight. This simple change destabilized the entire country, and it still has yet to recover. No amount of "cheap" electricity could make up for the cost of downtime, repairs, or lost business. In the end, it almost destroyed the operation.

A stable, resilient grid isn't a luxury. It's the foundation of prosperity, safety, and trust in any community. Anyone who ignores the risk of blackouts or brownouts, or thinks you can fix energy by simply moving jobs around without a developed plan, is gambling with people's jobs, health, and future.

WHO REALLY PAYS FOR ENERGY? (HINT: IT'S ALWAYS THE CONSUMER)

Politicians continually promote an energy policy myth. They would like us to believe that businesses will absorb the costs of carbon taxes, green mandates, or the next layer of regulation. It would be handy if government programs or corporate profits shielded regular people from rising energy bills, but the numbers and your own wallet tell a different story.

Let's be honest. Everyone knows businesses have to make money. When energy costs rise, they have to pass along those costs. Even the most community-minded corporations can't afford to lose money. They mark up prices, trim jobs, and do whatever it takes to survive.

Some people buy into the myth of government subsidies saving us money. But government money is just taxpayer dollars rerouted. No matter how you slice it, "free" programs are funded by people like you, through taxes now or debts you (or your kids) will pay later.

Here's what the data actually shows:
- 80% of energy cost increases are ultimately passed directly to consumers. Only a fraction is truly "absorbed" by business. Families, workers, and the middle class pay most of this 80%.
- Residential customers make up the single largest users of electricity in the US (about 37%).
- When you break down the statistics on household electricity bill payees, the burden falls hardest on the middle class and working families.

**Energy Cost Increases
Who Actually Pays?**

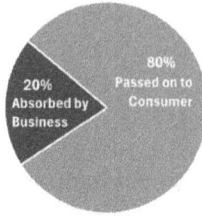

80%
Passed on to
Consumer

20%
Absorbed by
Business

**Who Uses Electricity
in the U.S.?**

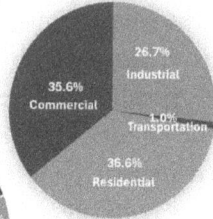

26.7%
Industrial

35.6%
Commercial

1.0%
Transportation

36.6%
Residential

**Who Pays Household
Electric Bills?**

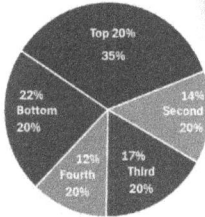

Top 20%
35%

22%
Bottom
20%

14%
Second
20%

12%
Fourth
20%

17%
Third
20%

WHY THIS MATTERS

Rising energy costs squeeze the middle class, shrinking their ability to spend, save, invest, or enjoy life. It hurts small businesses and local employers who can't pass along every cost. It quietly reduces the quality of life for everyone who isn't wealthy or subsidized. Subsidies and political gimmicks might delay the pain, but they never erase it.

It's important to remember two things. The "government" doesn't pay for anything. YOU DO!! "Big business" doesn't absorb costs; they pass them on to you at the cash register, on your bills, or in lost opportunities. That's why real cost transparency and genuine competition in energy aren't just economic issues. These essentials protect the backbone

REAL COST TRANSPARENCY AND GENUINE COMPETITION IN ENERGY AREN'T JUST ECONOMIC ISSUES.

of America: the middle class. Transparency and competition ensure that families, communities, and small businesses have a future that's not priced out by bad policy or bureaucratic fantasy.

SUBSIDIES: NOT THE ENEMY, BUT A TOOL

I don't want you to think I am anti-subsidies. Subsidies have proven essential to America's energy prosperity, and they'll continue to play a role when used wisely. In the 1950s and '60s, government investments in research, demonstration plants, and enrichment made civilian nuclear power possible. In the 1980s and early 2000s, the Department of Energy spent funds for research and development in the area of shale gas. That, and the tax breaks for horizontal drilling, helped unlock vast new resources that made America an energy leader. And in the last twenty years, tax credits for wind and solar power have driven costs down by 70-90% and jump-started entire industries. These subsidies worked because they followed a few common-sense guidelines:

- They target the "valley of death"—Promising technologies need help bridging the gap from lab to market.
- They are performance-based—Subsidies are most successful when politics don't play a role in their award.
- They share the early risk—Private investors need to see a bit of light at the end of the tunnel before they put their money in. However, as technology matures, subsidies should be withdrawn, allowing private capital to do the heavy lifting.
- Subsidy support should taper and phase out as costs fall.

- Subsidy support should be cut off if there are no signs that the technology will become less expensive or grow after ten to fifteen years.

The bottom line is subsidies should prime the pump, not become a permanent crutch.

LEARNING FROM FAILURES

These stories of blackouts, South Africa's rolling power failures, and what happened to rural towns in the 1983 freeze all offer the same lesson. When you underinvest in reliability, you ultimately pay more. When you ignore local expertise, you create problems no spreadsheet can predict. And when you chase cheap, you trade away the future for a quick win.

Whether it's through higher bills, lost jobs, or missed opportunities, Entrepreneurs, workers, and families who keep America running always pay the price. That's why L&H and Evercore hold every decision to a higher standard, and why we'll never accept policies that simply pass the buck to those who can least afford it.

The Bottom Line:
Every time our companies make an energy
determination, we ask three questions:
- Would I bet my family's business on this
recommendation? -
- Do I want my name on it for decades to come? -
- Can I trust it to keep a mining crew safe,
a hospital open, a small town alive? -
If the answer to even one question is no,
we start over.
Because at the end of the day,
the real price of energy
is paid by the people who trust us.
I won't let them down.

CHAPTER 3
HOW DID WE GET HERE?

Energy has been the cornerstone of every civilized society, from fire's first warmth to today's power grids. Every leap forward—whether it was windmills on the Nile, wooden pumps on the plains, or coal-fired steam engines—was driven by innovation, necessity, and the stubborn will to build a better life.

Early Americans powered mills by harnessing rivers. The industrial revolution was built on steam, coal, and wood, fuels that turned cities into economic engines. Nobody worried about carbon, and nobody—certainly not Washington—tried to regulate what kind of wood or coal you burned to heat and feed your family. They used a simple threefold measure. Does it work? Is it worth the effort? Does it last?

When nuclear energy arrived, the military proved its value first. Microreactors powered bases from Wyoming to the South Pole, surviving years of temperatures as low as -50°F without refueling. That same pioneering spirit lives on today at L&H and Evercore. We embrace the legacy my father-in-law, Jerry Schloredt, left when he ran those reactors for the Navy in some of the harshest environments on Earth. We build what works and never gamble with safety.

A Balanced Approach—Not Political Hype

After thousands of years of taking wood, coal, and oil from the earth, we understand the value of "giving back." But panic and fads won't get it done. I've seen too many leaders swept up in alarmist headlines or the "energy cure of the week." Many don't even look at the science behind the theories. They just latch onto whatever feels right or gets them applause at the next conference.

> NUCLEAR ENERGY OUTPERFORMS IN SAFETY, QUALITY, AND COMPETITIVENESS

At L&H and Evercore, we want cleaner, better energy. We advocate for nuclear because it's a great way to deliver on all Three Drivers at scale for real customers, not because we're climate activists. Nuclear energy outperforms in all three areas:

- **Safe:** The safety record of civilian nuclear energy, especially microreactors, is unmatched. Early military accidents were tragic, but rare, and provided lessons we still use today.

- **Quality:** No other technology can run for years without refueling, handle extreme environments, or offer the same uptime.

- **Competitive:** With modern innovation and fewer bureaucratic obstacles, nuclear can be more cost-competitive than ever, especially when you factor in reliability.

We're not writing off renewables, but we want to be honest about them. Wind, solar, and hydro have their place, but they bring environmental trade-offs, cost realities, and security risks. Imagine running a factory, hospital, or city on batteries alone during a polar vortex. When policy ignores these facts, the working class and industry will pay the price.

America's centralized grid has offered affordable, reliable energy to cities and factories for years. But when a hurricane hits, or one line fails, millions lose power, and rural or industrial communities get left out in the cold. The centralized grid is efficient, but expensive to maintain. And expansion to areas with lower population density is impossible. A better future means blending the best of centralized efficiency with local, entrepreneurial solutions like microgrids, modular reactors, and co-op power led by entrepreneurs that put customers, not bureaucrats, in control.

Entrepreneurs, rather than regulators, have driven every energy transition in history. But for decades, well-meaning policies have created perverse incentives, higher costs, and hidden risks like:

- Overregulation that stifles innovation.
- Subsidies that sometimes last too long, turning helpful bridges into permanent crutches.
- "Buy-back" rules and mandates that can make electricity unaffordable for the people who need it most.

Those who write the rules can not also regulate accountability. To be most effective, responsibility should always flow to those who pay the bills and take the risk.

WHAT IF WE STOPPED USING FOSSIL FUELS TOMORROW?

Everywhere you look, someone's calling for an end to coal, oil, and gas. But the science and common sense say otherwise.

- You can't feed cities without gas for farm tractors or oil for transport.

- Renewables can't cover peak demand during extreme events, and the batteries and minerals they require are produced overseas with fossil fuels.
- Shutting down fossil fuels overnight would collapse food supply chains and trigger rolling blackouts across the country.

We don't worship carbon, but we won't lie about how much our economy depends on it or pretend we can live without it right now. We need every energy source—renewables, fossil fuels, the grid, and nuclear—as we continue to add every resource we can find.

The American energy journey is full of brilliant leaps and frustrating failures. The difference, every time, is leadership. In addition to planners, we need builders and people willing to put their name and business on the line, not just their opinion.

> WE NEED EVERY ENERGY SOURCE—RENEWABLES, FOSSIL FUELS, THE GRID, AND NUCLEAR

At L&H and Evercore, we want to bring back that original spirit of accountability. We don't accept any suggestion blindly. We judge every energy solution by the Three Drivers: Safe, Quality, Competitive.

The Bottom Line:
If it doesn't serve people first, it doesn't get built.
If it doesn't work when it's needed most,
it doesn't get a pass.
If it's not something I'd put my family's name on,
it's not ready for America or its allies.
Those standards have brought L&H and Evercore
successfully to this point,
and that's how we'll move forward.

Chapter 4

The Energy Industry Today and Current Industrial Challenges

The truth is simple: if you run the world's biggest machines, "net zero" is a fantasy unless you include nuclear or some kind of breakthrough innovation. Theory will not help L&H's customers. They need energy that's Safe, Quality, and Competitive. The heavy industrial sector simply has no non-fossil option today that meets all three.

It's also time to remember the first principle of energy—energy cannot be created or destroyed, only transformed. Science tells us that ALL ENERGY is renewable, so the real question isn't whether fossil fuels are "dirty" or renewables "pure." We need to ask how we can transform energy in a way that actually works for the people who build, produce, and keep our country running.

Centralized Energy Limits Revealed

The big promise of centralization—efficiency, reliability, "one size fits all"—is falling apart. As I mentioned, it works fine until disaster strikes or policy fumbles. Take the 2025 blackout in the Iberian Peninsula.

Spain and Portugal, world leaders in wind and solar, received 80% of their power from renewables. But one grid overload resulted in 55 million people losing power for ten hours. Experts blamed weak interconnections and the intermittent nature of renewables, but the root cause was over-centralization. A distributed grid would have contained the outage and protected more communities. We've seen the same story in Florida, Italy, and England. With everyone connected to the same central source, a single event becomes a national crisis.

Unreliable grids, higher costs, and more regulations have forced industrial businesses (hospitals, mines, even schools) to begin to build their own micro-utilities. One survey estimated that 44% of businesses were considering a microgrid for their organization.[2] These entities are not energy experts. They're doing this just to keep the doors open. Some opted for micro-utilities to cut costs; others want certainty against blackouts, brownouts, and wildfires. But what they mean for these businesses is that instead of being able to focus on what they do best, companies have become energy managers by necessity.

THE PROBLEM WITH POLITICS AND OVERREGULATION

The biggest risk to America's grid is no longer technical; it's political.

> THE BIGGEST RISK TO AMERICA'S GRID IS NO LONGER TECHNICAL; IT'S POLITICAL.

Think about a football game. What would happen if they put one referee per player on the field? We wouldn't have any plays, just endless whistles.

The more "referees" you put on the field, the more the game grinds to a halt.

That's today's energy market.

There are times when we expect the government to act. We need it for defense, highways, and public safety. But when every decision needs political sign-off, innovation dies. To set productivity free, the private sector needs to be free to solve smaller-scale problems while the government focuses on keeping the largest projects and bad actors in check.

DECENTRALIZATION: THE REAL PATH FORWARD

True resilience and competitive costs come from distributed, decentralized energy. We've already seen this play out in telecom and space.

- **Telecom**: AT&T's one-hundred-year monopoly kept phones expensive and tech stagnant. When the industry was broken up and deregulated in the 1980s, prices plummeted, and new technology exploded.

- **Space:** NASA ruled for decades, but after SpaceX and private partners entered the arena, launch costs dropped 99% and reusable rockets became a reality.

Evercore and L&H want to bring that model to the energy sector. Private leadership, competition, and entrepreneurial speed have the potential to revolutionize the energy industry. We're ready to be the Uber® of energy, matching customers with proven microreactors and energy innovators, so industries can focus on production, not bureaucracy.

Of course, even in a decentralized system, the government still has a vital role. They provide guardrails for bad actors. We need them to set high standards for gigawatt-scale plants and public safety and fund early-stage research in the labs, especially for nuclear innovations and microgrids. Government subsidies will also provide a tool for innovators to cross the "valley of death." At the same time, it's imperative they don't become a crutch. They have to step back and let the private capital work.

GOVERNMENT SUBSIDIES SHOULD BE A TOOL, NOT A CRUTCH.

But until decentralization occurs, overregulation will put the free market in handcuffs. Today, the rules could keep a Wyoming data center from receiving an Evercore microreactor, even if it solves the problem, delivers safety, and brings competition. Local utilities and state-by-state rules keep private solutions locked out, especially for smaller customers who need it most.

ENTREPRENEURIAL LEADERSHIP IS THE MISSING LINK

The clean energy debate doesn't need more politicians, bureaucrats, or regulation. What we're lacking is entrepreneurial leadership—people who build, innovate, and take ownership. When leaders are intentional, set clear mission and values, and focus on legacy, you get real progress without waiting for the next government handout or act of Congress. For instance, Wyoming's mining industry reclaimed and restored its own lands because it cared about the future and local accountability, not because a law forced it to.

Evercore plans to be an entrepreneurial leader in changing the way our nation harnesses energy. Our business model is simple:

- Buy and operate microreactors as a service for heavy industry.

- Let the customer focus on their core business, while we deliver Safe, Quality, and Competitive energy.

- Repeat, scale, and keep costs down—just like telecom, just like SpaceX.

Bad actors will always exist. But the answer isn't more red tape for everyone. We need clear standards, real penalties, and permission for the market to choose proven solutions. The majority want to do what's right for the land, for people, and for the next generation if the system lets them.

The world envies America's seventeen national labs and research infrastructure. All we need now is a better partnership between private industry and the government. If private leaders scale new technologies, while the government focuses on safety, research, and smart subsidies, America can avoid letting short political cycles dictate long-term investments and keep a tight focus on public value, rather than special interests.

Transportation and telecommunications prove this approach works. With a strong partnership that puts our energy future in the hands of the private sector, competition and innovation will be unleashed at the ground level, and we'll begin to see higher quality and more affordable energy.

The Bottom Line:
It's time for an energy policy that rewards building,
not just talking.
If we empower entrepreneurs and let them lead,
the next era of American energy abundance
will belong to everyone.

PART II

Innovation, Leadership, and the New Energy Playbook

CHAPTER 5

TECHNOLOGY-DRIVEN TRANSFORMATION

L istening to mainstream media, you'd think nuclear energy is stuck in the past, defined by three old disasters, not new breakthroughs. But that story is out of date. The truth is, engineering and business innovation have changed the game, and the leaders who adapt fastest will shape the future of energy.

NUCLEAR: SAFER, SIMPLER, AND MORE COMPETITIVE THAN EVER

When you hear us boast that nuclear is safer, we're really saying something. Since the beginning of nuclear energy, only solar has experienced fewer accidents.[3] It's been the safest energy source for decades, and now it's even safer.

Nuclear power isn't magic; it's physics. When the nucleus of a uranium atom splits (fission), it releases a burst of energy. That energy is captured by a coolant, converted to heat, and used to make electricity. For decades, the process was the same: big, expensive, and overengineered for a level of risk that modern science has since beaten.

Today's Advanced Nuclear Reactors are smaller, safer, and dramatically more efficient. Standardization and modular designs, already proven in places like France, slash costs and make safety more manageable.

- 95%+ uptime is now standard.
- Higher operating temperatures mean nuclear energy can power heavy industry—not just the grid.
- Nuclear now offers less waste, less risk, and more value.

Death rates per unit of electricity production

Our World in Data

Death rates are measured based on deaths from accidents and air pollution per terawatt-hour[1] of electricity.

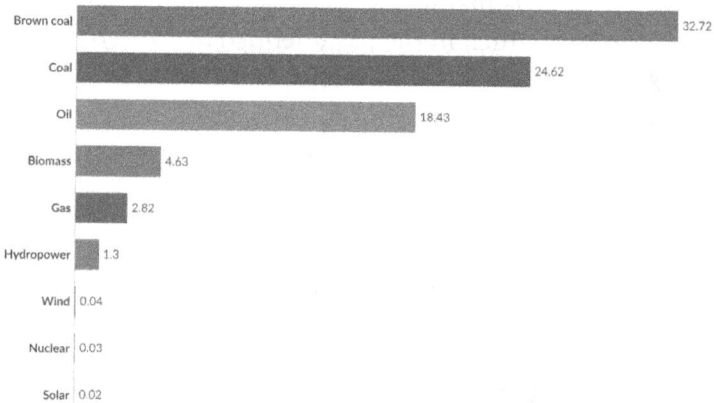

Brown coal	32.72
Coal	24.62
Oil	18.43
Biomass	4.63
Gas	2.82
Hydropower	1.3
Wind	0.04
Nuclear	0.03
Solar	0.02

Data source: Markandya & Wilkinson (2007); Sovacool et al. (2016); UNSCEAR (2008, & 2018)   OurWorldinData.org/energy | CC BY

1. Watt-hour A watt-hour is the energy one watt of power delivers for one hour. Since one watt equals one joule per second, a watt-hour equals 3600 joules of energy.
Metric prefixes are used for multiples of the unit, usually:
- kilowatt-hours (kWh), or a thousand watt-hours;
- Megawatt-hours (MWh), or a million watt-hours;
- Gigawatt-hours (GWh), or a billion watt-hours;
- Terawatt-hours (TWh), or a trillion watt-hours.

For years, the media has broadcast the world's nuclear failures. However, while they focused on accidents that produced far fewer fatalities than coal, oil, or hydro, the US military quietly ran microreactors under extreme conditions. In places like Greenland, Antarctica, and Panama, they powered bases for years without refueling, breaking safety records the news never bothered to mention.

> ADVANCED NUCLEAR REACTORS ARE SMALLER, SAFER, AND DRAMATICALLY MORE EFFICIENT.

THE NEXT-GEN FUELS: HALEU AND TRISO

But nuclear has more than smarter reactors to boast about. The fuel itself is better than ever. Today's advanced nuclear reactors use new fuels such as HALEU (High-Assay Low-Enriched Uranium) and TRISO (Tristructural Isotropic). These raise the bar for safety, efficiency, and real-world performance.

Here are the quick facts about these new high-quality fuels:

- HALEU is uranium enriched to 5–20% U-235. Conventional reactors run on uranium enriched to 3-5% U-235.

- HALEU delivers more power from less material, allowing for smaller, modular reactors and reducing the number of refueling outages.

- Most microreactor and next-gen nuclear designs require HALEU. That's why the US and its allies are investing heavily in HALEU supply chains now.

- Each tiny TRISO fuel kernel starts with HALEU at the core, then is wrapped in three tough layers of ceramic and carbon, making each particle its own virtually indestructible containment vessel.

- Even if a reactor is breached, each TRISO particle is its own mini-barrier—no meltdown, no major leak, and unmatched safety.
- TRISO handles much higher temperatures than conventional fuel, nearly eliminating the "worst-case scenarios" that haunted old-school nuclear.
- TRISO enables radically safer, modular, and distributed reactors.

BWXT (BWX Technologies) is the US leader in producing both HALEU and TRISO fuel at commercial scale. Companies like X-energy and others are now proving TRISO-powered microreactors in the real world.

WHAT'S REALLY HOLDING BACK NUCLEAR?

Alright. We have the fuels. We have the blueprints. It's not physics keeping us from manufacturing microreactors. Maybe you're wondering, *with all these pieces in place, what's the holdup?*

Here's the truth: advanced fuels and clever reactor designs won't mean a thing if we can't build them. The shop floor is where safety and quality are made real, not in the PowerPoint presentations or the permits. Nuclear's bottleneck isn't physics—it's manufacturing. And manufacturing has been halted by paperwork and outdated policy.

The Nuclear Regulatory Commission still uses standards based on 1970s technology, and every small change requires years of review and millions of dollars in compliance costs. Right now, we have overengineered designs and endless "custom" specs that slow everything down.

France fixed this by building standardized fleets, and when it did, costs plummeted. In America, most "safety" regulations add time and bureaucracy, not actual safety. If we want real progress, we need a regulatory system that encourages innovation, rewards real safety, and holds everyone accountable.

The safety record of nuclear energy isn't an accident. It is a result of relentless discipline—weld procedures qualified, every heat number tracked, every part tested and retested. Now it's time to turn the process over to private sector companies with a proven track record.

L&H has spent sixty-one years living in that reality. My father built the business by taking over the mega machines the original manufacturers abandoned—the NASA crawler transporter,

> THE SAFETY RECORD OF NUCLEAR ENERGY ISN'T AN ACCIDENT. IT IS A RESULT OF RELENTLESS DISCIPLINE.

rope shovels, mineral processors. Those machines only came to us when they were broken, and downtime cost millions. There was no one else left to call, so L&H used that pressure to forge our culture: rebuild it, upgrade it, and make it run safer and longer than before.

That cycle of failure, learning, redesign, and rebuild safer is exactly what nuclear needs as it restarts. L&H understands that discipline. It's the world we live in every day. And the reality of what this kind of discipline can produce is why I believe America can make microreactors not just possible, but practical if we put manufacturing back at the center.

Anyone can raise capital or publish a white paper. The companies that will last are the ones with people who *love* making things, who grind through downturns, and who never compromise safety or quality. That's the type of culture nuclear demands. That's what will separate the flash-in-the-pan start-ups from the teams who will still be standing when the first fleets of microreactors need servicing thirty years from now.

AI: THE ULTIMATE ACCELERATOR FOR ENERGY INNOVATION

The next leap won't just come from new reactors. It will depend on how fast we can put them to work. At Evercore, we closely track BWXT and X-energy's progress. When HALEU and TRISO-powered microreactors are ready for energy customers, and there's real demand to aggregate, we'll use the latest AI technology to compile the tools that will connect customers with the right solution. We will deliver safe, high-quality, and competitive energy in places that have never had it before.

AI can cut through the red tape, simplify permitting and compliance, and rapidly match microreactor supply with real customer demand. This will mean lower costs, faster timelines, and better risk management for everyone involved. Most importantly, AI will free up human expertise for education, safety, and customer support—rather than burying people in paperwork. The results will be staggering:

- Projects that used to take a year can be launched in weeks.
- Lower cost means more businesses can compete, and the middle class isn't stuck holding the bill.
- AI unlocks data-driven decisions, not political guesswork.

BUSINESS MODEL: PRIVATE LEADERSHIP MEETS NEXT-GEN TECH

Evercore's approach is simple. We will own and supply advanced microreactors, delivering clean, safe, and reliable energy to heavy industry on demand, as a service. Then, as microreactors and market demand mature, our AI system will help us aggregate and serve customers faster and more efficiently than ever.

L&H will provide parts for reactor companies to bring old systems back online, light them up with our Operational Intelligence Platform (OPS IQ) digital Internet of Things (IoT) system and modern sensors and controls, and deploy OPS IQ Mentor. The AI knowledge aggregator will use the digital data being gathered, along with the knowledge and experience of all operators and maintainers, to train the next generation and perpetually improve the team's performance as they keep the machines running.

L&H will focus on building supply chains for microreactors and other nuclear companies. At the same time, Evercore will match the winning microreactors and possibly other innovative energy solutions to our customers in mining, mineral processing, and other heavy industrial or national security applications. These endeavors don't usually have the luxury of tapping into a quality grid with enough capacity to serve them. L&H has always served and innovated for energy companies, but now we are taking it up a few notches.

We refuse to wait for someone else to fix the system. We're building the playbook for a new era. SpaceX rewrote the rules for space, and we're finding the places where our unique capabilities and passions make compelling offers for customers that share our mission, vision, and values.

KEEP IT SIMPLE AND ALWAYS INNOVATE

One of the best things we can do to access energy abundance is to simplify product development. Elon Musk's Five-Step Design Process empowered SpaceX to drive down costs and speed up safe, quality production.[4] These five steps have the potential to take us to an energy win:

1. **Question every requirement.** Most rules aren't as necessary as they seem.

2. **Delete what you don't need.** Simplicity beats complexity every time.

3. **Only then, optimize.** Never optimize what should have been deleted.

4. **Accelerate after optimizing.** Move fast but only in the right direction.

5. **Automate the proven system.** That's when you scale, and AI becomes your superpower.

L&H and Evercore will always keep innovating, pushing for smarter regulations, and putting the needs of real people first. Because if we don't, no amount of tech will matter.

The Bottom Line:
Technology is here to serve people.
An excellent culture built to deploy technology in flow
is the superpower combination.
With the right leadership,
the next wave of energy abundance is
closer than most believe.

CHAPTER 6

THE OWNER CO-OP MODEL—ENTREPRENEURS, INCUMBENTS, AND SERVING THE RIGHT CUSTOMER

When the government passed the Rural Electrification Act in 1935, it was about more than lightbulbs. They gave power—literal and economic—to people who'd been left out. In the early days, the big private utilities ignored rural America, seeing nothing but cost. Roosevelt's New Deal didn't just build dams—it enabled communities to form co-ops, pool risk, and finally turn the lights on for the farms and towns that industry overlooked. Within a generation, ninety percent of Midwest farms had electricity—thanks to local ownership, shared risk, and a federal push to empower the little guy.

HOW CENTRALIZATION BUILT—AND STALLED—THE GRID

Today, most US electric providers are still rural co-ops. Decades of regulation and centralization changed the original vision. What started as a plan to provide loans to groups of farmers to bring power to rural America has turned into:

- The biggest co-ops and investor-owned utilities receiving legal protection and guaranteed profits.
- A roadblock for innovation and building new electric generation without their permission in many territories.
- Reliance on aging, often coal-based, power plants by small co-ops that distribute but don't produce or transmit energy.

This model made sense when transmission was expensive and reliability meant size. With cost and reliability still at the forefront, today, the real issue isn't hitting "net zero." It's about changing the model to provide even more reliable service at a price everyone can afford while giving communities more control over their future.

The protected incumbents seem to slow-walk change. They know any major upgrade—whether for cleaner energy or simple modernization—ends up being paid for by voters and taxpayers, not by the utility itself. And as public focus shifts from emissions to the reality of higher bills and reliability risks, pressure builds for a new, more flexible model.

One of the reasons the grid has stalled progress is because it treats every customer exactly the same. However, every region, industry, and business brings unique needs to the table. Some put lowest cost above all else. Others prioritize resilience, reliability, and control, especially mission-critical or remote operations.

> EVERY MAJOR UPGRADE ENDS UP BEING PAID FOR BY VOTERS AND TAXPAYERS.

For many urban and suburban businesses, a strong, reliable, working grid will be the first and best choice. If it isn't, someone is not doing their job. There will always be room for large incumbents, regional co-ops, and entrepreneurial disruptors. The key is understanding who you're serving and making sure each customer has the right solution for their specific needs.

EVERCORE'S FOCUS: WHERE THE GRID ISN'T AN OPTION

At Evercore, we know we can't be everything to everyone. We have a clear mission: serve the world's most remote, energy-intensive, and critical operations. We focus on mines, mineral processing plants, heavy industry, and national security projects, places where grid power isn't an option.

For these customers, self-reliance, security, and up-time aren't just nice to have; they're a matter of survival. If you have to choose between the grid and a microreactor, and the grid is strong, you should use the grid.

However, if price, politics, or unreliability get in the way, you should have other choices. That's where entrepreneurial solutions step up. Most people don't know the planning that goes into placing a mine or mineral processing equipment. "Can we rig up and scale down the energy we need?" becomes a pivotal factor and drives costs.

When we remove the grid reliance component, suddenly, the position of the mine and processing is based on the deposit rather than logistics. This changes the world in a positive way.

> BEING THE LEADER IN ENERGY INNOVATION IS A MATTER OF NATIONAL SECURITY.

The countries that lead the separation from the grid need to be the USA or its allies. I prefer to see the USA lead the nuclear microreactor deployment for the free world. Being the leader in energy innovation is a matter of national security. A world full of microreactors owned and operated by Russia or China is not ideal for the free world, and it's not as safe.

A FLEXIBLE FUTURE—PARTNERSHIPS, COMPETITION, AND REAL VALUE

The future of energy isn't about one system over all others. It's about partnership, local knowledge, innovation, and making sure the right model and the most appropriate energy sources are applied to the specific challenge. We can create a scenario where everyone wins:

- Utilities, co-ops, and government win when they continue to provide service in high-density, low-risk areas.
- Entrepreneurs, innovators, and tech-driven providers win when the government loosens the reins and lets them take the lead in hard-to-serve, remote, and critical applications.
- Customers win when they get Safe, Quality, Competitive energy tailored to their real-world situation.

The Bottom Line:
Real energy abundance comes
when customers are served where they are,
not forced into one-size-fits-all models.
The leaders who thrive will be the ones
who listen, adapt, and deliver the best value
locally, globally, and everywhere in between.

CHAPTER 7

NUCLEAR ENERGY AND MICROREACTORS TRUTH, INNOVATION, AND THE REAL WORLD

Nuclear energy gets a bad rap, primarily because of fear, politics, and leadership breakdowns rather than facts. If you want the real story, you have to look beyond the headlines and find out what actually happened, what we learned, and where innovation is taking us now.

THREE MILE ISLAND—A LESSON IN LEADERSHIP FAILURE

The partial meltdown at Three Mile Island in 1979 was not just a technical incident. It was a crisis of leadership, communication, and trust. The technical failure started with a non-nuclear part of the plant, compounded by faulty gauges. But the real damage came from poor crisis management. A lack of clear, honest communication eroded public trust and fueled unnecessary panic. In the end, the *actual* physical impact was minimal. They had no detectable health effects for workers or the public. However, the incident set back public confidence in nuclear power for a generation.

CHERNOBYL, FUKUSHIMA, AND UNDERSTANDING ENERGY RISK

Chernobyl was a different story—no containment, poor design, compounded by leadership failures, cost lives, and left a legacy of fear. Fukushima was a worst-case disaster, compounded by human mistakes. But each failure made nuclear safer by forcing hard lessons and better standards.

THE REAL SAFETY NUMBERS OF NUCLEAR

Every energy source has risks, but nuclear stands apart for its safety record—especially in the US.

- In the United States, nuclear power has not caused a single death from radiation exposure—despite more than sixty years of continuous operation and some high-profile incidents.
- Other energy sources, like coal and oil, have caused thousands of deaths annually from air pollution, mining accidents, and respiratory illness.
- Hydroelectric disasters—like dam failures—have killed thousands in single events.
- Even wind and solar, considered "safe," involve worker fatalities. When compared to nuclear, wind has more fatalities per unit of electricity produced, and solar falls just slightly lower, mainly due to construction, installation, and maintenance risks.

Worldwide, there have been a few tragic nuclear incidents, but even when including those, nuclear remains the lowest for fatalities per terawatt hour produced—by a wide margin.[5]

> NUCLEAR REMAINS THE LOWEST FOR FATALITIES PER TERAWATT HOUR PRODUCED

RADIATION: ESSENTIAL TO LIFE, POWERFUL IN MEDICINE, SAFE WHEN MANAGED

When someone mentions radiation, people get nervous. They picture those heavy aprons we wear during X-rays or remember the last radiation-ends-the-world movie they watched. But there's an important truth we need to start spreading: not all radiation is dangerous. In fact, radiation is a fundamental part of life on Earth.

- **The sun's radiation** powers photosynthesis, drives the water cycle, and provides the vitamin D every human needs to survive.

- **Low levels of natural background radiation** are everywhere—soil, rocks, even the food we eat. Life evolved with it, and plants, animals, and people all depend on it in small, natural amounts.

- **Radiation is an important partner with medicine.** This is what sets nuclear energy apart from all the others. Nuclear reactors don't just make electricity; they produce medical isotopes that save millions of lives every year. These isotopes, like technetium-99m, are harvested from reactors and used for imaging, cancer treatment, and diagnostic tests worldwide.

Here's the irony: if a tiny amount of that harvested radio-active isotope spills between the reactor and the packaging room, it's a full-blown safety crisis. You have to implement special procedures and call in trained teams. All kinds of unnecessary regulations kick in. But if you take that same isotope, change the label, safely package it, and mail it to a doctor, the risk changes to "normal." That doctor will inject it straight into a patient or use it to create scans. Afterward, it will pass through the body and be flushed down the drain as routine hospital wastewater without any special concern.

With radiation, context and control are everything. There are dangerous types and doses of radiation, safe and essential ones, and everything in between. In industry and medi-cine, the people who handle radioactive materials must be well-trained, disciplined, and serious about safety. And for more than seventy-five years, American nuclear professionals have proven just how well that works.

RADIATION TYPES

I think most people would feel better about nuclear energy and radiation if they better understood the various types. So here's a quick breakdown of the various types of radiation from least to most hazardous in typical nuclear energy and life scenarios:

- **Alpha Radiation (α)** is only dangerous if you inhale or ingest it. The radiation particles are large, so your skin or a sheet of paper will stop them.

- **Beta Radiation (β)** consists of electron-sized par-ticles that can be easily stopped by plastic, glass, or aluminum. It can be more dangerous if it's ingested and offers a slight risk to exposed skin.

- **Gamma Radiation (γ)** is used for super-powered X-rays and can be stopped by concrete or lead. Because it can pass through the body, shielding is critical in reactors and medical use. Typical X-rays produce radiation from outside the nucleus of the electron.
- **Neutron Radiation** is not present outside reactors under normal conditions. Concrete and water stop it.

THE NEXT WAVE: ADVANCED NUCLEAR AND MICROREACTORS

The future of nuclear power isn't just massive plants locked to the grid. It's a mix of all advanced nuclear technologies—including large-scale reactors, next-gen designs, small modular reactors (SMR), and, most exciting for our Evercore Energy vision, portable microreactors built for industrial, national security, and other remote applications.

Evercore wants to take the lead in powering up those locations that work most efficiently far from the grid. We want to take energy concerns out of the equation when they begin laying out the framework for their most advantageous locations. Microreactors answer this need in three specific ways:

> THE FUTURE OF NUCLEAR IS A MIX OF ALL ADVANCED NUCLEAR TECHNOLOGIES.

- **Portability:** Microreactors are designed to be delivered, installed, and redeployed as needed, making them ideal for mines, remote towns, or field operations.
- **Up-time:** All advanced nuclear plants offer high capacity factors and long operating periods between refuelings.

- **Safety:** With innovations like TRISO and HALEU fuels, modern reactors can withstand scenarios that would have been catastrophic a generation ago.

NUCLEAR WASTE: GET THE REAL NUMBERS

Next in line after the fear of radiation, we hear the hype about nuclear waste. Here's the honest math:

- The nuclear waste the average American will leave after a lifetime of nuclear-powered electricity use amounts to about the size of the end of your finger. Some industry sources use a soda can as a visual—either way, it's a *tiny* fraction compared to coal, gas, or even solar waste per person.
- All US commercial nuclear waste ever produced would fill a single football field about thirty feet deep.
- New fuels like HALEU and TRISO make that waste even safer, more stable, and easier to store.

ALL ENERGY IS NUCLEAR ENERGY—IT'S JUST A MATTER OF DENSITY

Here's a truth you won't often hear: all energy on Earth is ultimately nuclear energy. The sun is a giant nuclear fusion reactor, and its energy drives wind, solar, and hydro. Even trees, coal, and biomass we've burnt for centuries started with photosynthesis, nuclear energy from the sun.

But nuclear fission brings a game-changer: **energy density.** A single uranium fuel pellet (about the size of a fingertip) contains as much usable energy as a ton of coal, 149 gallons of oil, or 17,000 cubic feet of natural gas. That's why nuclear plants need far less fuel and create far less waste by volume than any fossil or renewable source.

This means less mining, less manufacturing, less shipping, and far less waste per unit of energy delivered. Plus, the risks are lower in the fuel supply chain. Less exposure and fewer shipments equal a smaller environmental footprint.

Energy Source Comparison by Key Metrics

Energy Source	LCOE ($/MWh)	Capacity Factor (%)	Land Use (km²/ TWh)	Lifecycle Emissions (gCO₂eq/ kWh)	Grid Reliability Impact
Advanced Nuclear	$70-100	90-95%	1-4 km²	5-10	High (baseload)
Traditional Nuclear	$90-140	90-93%	1-4 km²	12-15	High (baseload)
Natural Gas (CCGT)	$45-75	50-60%	2-5 km²	400-500	Medium-High
Coal	$65-150	40-60%	8-10 km²	750-1000	High (baseload)
Solar PV (Utility)	$30-60	20-30%	40-50 km²	40-50	Low (intermittent)
Wind (Onshore)	$30-60	30-45%	50-70 km²	10-15	Low (intermittent)
Hydroelectric	$30-200	40-60%	Varies widely	15-25	High (flexible)

Additional Nuclear Facts

- Solar energy emits more greenhouse gases than nuclear energy (on lifecycle analysis).

- Nuclear energy has the fewest fatalities per terawatt hour—see the chart in chapter five.

- Nuclear is the *only* energy source producing medical isotopes for life-saving healthcare.

- Nuclear energy's land use is a fraction of solar and wind—critical for preserving open space.

GET PAST THE HYPE

I am a fan of all energy, and at L&H, we work on innovating and maintaining machinery for coal, oil, gas, hydro, wind, and now nuclear. We need it all. Let's use AI's intelligence and perpetual energy innovations to continue delivering excellent energy to the USA and the free world.

Every technology has its "miracle claims." Battery-powered mines and fully renewable grids sound great at a trade show, but in the real world, they're not ready for the heavy lifting. If you want to power a mine, a base, or a remote town with true 24/7 reliability, advanced nuclear microreactors prove very interesting in many applications, and for Evercore's focus, movable microreactors will be the solution as soon as they're ready.

The Bottom Line:
Don't let politics or old headlines
make the decision for you.
The facts, the charts, and the performance
in the real world prove that
Nuclear—especially in its next wave—can deliver
Safe, Quality, and Competitive energy,
as well as a more abundant future for all.
AI fed with excellent, real-time, digital data
will help us analyze and apply energy solutions
that are competitive and best for the human race.

PART III

THE HUMAN FACTOR

CHAPTER 8

ENERGY INNOVATION
STARTS WITH YOU

When we talk about energy, minds go to electricity and heat. We imagine oil fields, coal mines, and reactors. But what if I told you I have discovered a more powerful source of energy? This resource is the most underutilized producer you will ever encounter. It's the Power Plant Between Your Ears. Fueled through mental fitness and flow, your brain is the Real Source of Abundance.

Before you can lead others, build the next great innovation, or transform an industry, you have to unlock your own capacity. No matter how advanced the technology, the ultimate bottleneck is always human leadership, creativity, and resilience.

In my journey at L&H and Evercore and in writing my first book, *Know Thyself, Love Thyself*, I learned that energy innovation doesn't start in the lab or the boardroom. It begins inside me.

Mental Fitness & Resilience—The Real Foundation of Flow

Mental fitness and resilience are the *real* foundation of flow. You can't reach peak performance—at work or anywhere else—if your mind is weighed down by stress, trauma, or old baggage. Until you build that foundation, flow will always be out of reach.

Work-life balance is a key part of the equation. For a person to be able to flow, they need to feel balanced—however that looks for them. And as we age, work-life balance is even more critical.

Great leaders don't assume everyone wants the same hours, schedule, or boundaries. They recognize that what balance is for one person is burnout or boredom for another. The only way to unlock a team's full energy potential is to understand and honor these differences and to make sure your team's actual life balance matches the job you're assigning them.

> WHEN YOU GENUINELY CARE ABOUT YOUR PEOPLE'S UNIQUE CAPABILITIES AND PASSIONS, YOU CREATE LOYALTY, LEGACY, AND LEGENDARY RESULTS.

Working with your team to discover and apply their strengths in a meaningful way better serves your customers. Plus, something amazing happens. You'll discover most people feel truly seen and loved for the first time ever. This means customers are dazzled by the performance of a person who operates in flow, showcasing their real, natural uniqueness with excellent mental fitness and resilience.

This is more than just good business. When you genuinely care about your people's unique capabilities and passions, you create loyalty, legacy, and legendary results.

HOW MUCH POWER IS YOUR BRAIN REALLY PUTTING OUT?

Here's a fact that changes everything: Your brain has around fifteen watts of power to work with. But most people deploy only two to three watts into productive thinking.

The rest gets robbed by:
- Past and present trauma—some you know about, some so rough your brain hides it until you're ready to handle it.
- Stress over relationships, kids, and family.
- Addictions and all the other mental junk we carry around.

When that stuff leaks out, you get triggered. You end up lashing out and overreacting, and it always happens at the worst possible time. Then you shove it down, hope it doesn't happen again, and just add another stone to the heavy bag you're already dragging.

Learn from my experience. Unpacking that bag is a hell of a lot easier than pretending it isn't there and towing it mile after mile. When you unpack it, you get more of your watts back for clear thinking, creativity, and your unique abilities and passions. That's when you start running closer to the horsepower you were built for.

WHY MOST TEAMS NEVER REACH FLOW

Here's the hard truth: Mental fitness is the hidden foundation of every high-performing team. Without it, people can't even see their mission, let alone

> MENTAL FITNESS IS THE HIDDEN FOUNDATION OF EVERY HIGH-PERFORMING TEAM.

reach "group flow." People stuck in survival mode, anxiety, or unresolved trauma never think about MVV (mission, vision, and values). They don't consider their Unique Ability and how they fit into the tribe.

The best leaders know this. They build **Mental Fitness First Teams**, and it shows up in every safety record, innovation, and lasting partnership. All elite special forces focus on mental fitness and resilience, not human strength. You know the saying, "It's 99% mental?" It is! What's it? Whatever you apply your time and ENERGY to.

FLOW: FROM INDIVIDUAL TO GROUP

Before you can experience group flow, you need a team that focuses on individual flow. When you're operating in your Unique Ability, time disappears, and you set records with ease. And when every person on your team experiences this heaven-on-earth state, productivity sours.

The Chicago Bulls of the '90s and world-class orchestras give us a glimpse of group flow. We experience it in the best field teams at L&H. Group flow takes us to the highest level of business or human performance. But to get there, each team needs a few basics:

1. A humble, persistent leader at the helm
2. Clear MVV
3. A team of people operating in their individual flow, aligned and pushing just beyond their comfort zone—about 5% beyond "easy"
4. Relentless focus on the team's state
   a. Herding cats when necessary
   b. Compensating for an individual's off-days

c. Never settling

d. Always making the hard decisions

e. Leading the team in service of the overall MVV

f. Constantly working for the good of all the tribes—employees, customers, vendors, and all other stakeholders in that order

"Flow" isn't magic. It's the result of intentional leadership, culture, and relentless clarity. And once you achieve it, it trickles through your organization—lifting every team, tribe, and customer experience. Once employees and customers experience flow, they won't want to leave it. Most find themselves sorely disappointed when they try something else or have to deal with other companies and vendors that aren't in flow.

HOW L&H TACKLED THE MENTAL FITNESS CHALLENGE

I used to think that if you hired the best and gave them great tools, culture would take care of itself. But the real secret is deeper: your people's brains come with their families, histories, and unresolved baggage. You can ignore that, but you will eventually pay for it in lost performance, missed potential, team breakdowns, and increased turnover.

When someone mentions Wyoming heavy industrial manufacturing and repair, mental fitness and resilience conversations aren't the first things you think of, even though they've become mainstream topics since COVID. At L&H, we have proven the benefits. Unfortunately, we've also discovered that it's a hard concept for companies to replicate even after we give them the resources. We try to help other businesses develop this culture because we know how crucial it is to innovation.

L&H's journey started with a chance meeting at "40 Years of Zen"—a neurofeedback program that claims to compress decades of meditation into a week. Everything changed when I met Travis Ramsey. His insights blew me away. Jerri and my sister Laura were with me, and immediately after Travis took me through the process, I knew everyone in my tribe had to have access to this power.

Normally, you find only successful executives at the 40 Years of Zen program. Three or four strangers come together for coaching. Bringing my wife and sister proved different and powerful. Travis helped all of us learn and improve in every way we wanted to at an incredible pace.

I knew I couldn't bring my entire family or crew to 40 Years of Zen, but I truly wanted to bring this life-changing experience home with me.

Travis started by coaching my family over Zoom. It worked so well that I asked him to work with my 250 employees. A few weeks later, he told me, "Mike, if I'm going to help your employees, I need to help their families. Almost all their issues start with their intimate relationships and kids."

It was a risk, but we did it. That's how the Honestly Better® Mental Fitness Program was born at L&H. We started with five coaching sessions a month and kept doubling. Today, Travis maxes out at about one hundred one-on-ones or group meetings a month, helping employees and their families with L&H footing the bill.

His work has:
- saved marriages
- kept employees from quitting because their lives were falling apart, and they were taking their careers with it
- helped their kids through serious crises

- tackled generational trauma, dealing with the hidden battles most people never see, much less have the resources to solve

I had no idea how distracted people are at work until I saw the value of helping my team resolve those distractions. The payoff is priceless. When people can focus on what's in front of them, you unlock more energy, more loyalty, and a workplace that truly thrives.

To take it one step further, we made Stacy Stuckey our internal Kolbe Gallup Clifton Strengths Coach. She works with our team to help them

> WHEN YOU CLEAR THE CLUTTER, PEOPLE GET THEIR BRAINPOWER BACK.

discover their unique abilities and ensure everyone is working in their strengths. This allows us to achieve flow on our own.

When you clear the clutter, people get their brainpower back. Suddenly, you see happier people who focus on safety, quality, and productivity. Retention and performance improve. Innovation and engagement follow. The benefits are immeasurable, which is exactly why most business leaders miss the point.

Since the benefits can't be tracked on a spreadsheet, it can feel like you're paying people to deal with family issues, not work issues. But if we don't help them fix their personal problems, our people can't focus on their professional lives. If they're worrying about their spouse or kids, that takes priority. The only way to get their focus back is to give them the best resources to resolve their issues.

This setup is unconventional and not licensed therapy. Our employees and their families can access that through EAP and/or insurance when it's needed. L&H Honestly Better Mental Fitness is more entrepreneurial and innovative and not boxed in by titles, licenses, or rules. Even with mental health, we go where success takes us. Working with our people, we iterate as needed and don't worry about measuring the results with KPI's and other statistics.

I check in once in a while to see what's working and what's not to make sure most of those served love it. But I believe this program works best when we let it run fast and loose. Not only is it results-oriented, but it is also personal. The conversations my team has with Travis are none of our business unless they choose to share. Our mental fitness coach and I are on the same page with privacy first.

MEET YODA, L&H INDUSTRIAL'S MENTAL FITNESS CHATBOT

Since we started the Honestly Better Mental Fitness Program, technology has made leaps forward. So, of course, L&H is taking advantage of it. People needed more Travis, but there was only one of him—and people have emergencies. So, Travis helped me build Yoda: L&H Industrial's Mental Fitness Chatbot on ChatGPT. Yoda has been loaded with the best teachers and thinkers from ancient wisdom to modern science, as well as all our Honestly Better Mental Fitness training.

We've openly shared the model, but so far, no other CEO has replicated it to scale. I hope this changes. The Honestly Better Mental Fitness model and Yoda work for us because of our unique combination of culture, entrepreneurial spirit, and MVV.

We're not trying to package and sell our model. Our sole focus is applying it to our tribes, and if it can benefit other businesses, so much the better.

That is what makes L&H, Evercore, and our tribes different. It's the same reason AI works so well for us; if it's not clear how it is helping, then we don't do it. Our model applies innovation to a person or group, with a clear expectation of performance gains. This makes it a sustainable and scalable innovation, and it applies to mental fitness and AI.

"INNOVATE ENERGY—STARTING WITH YOU" PLAYBOOK

At L&H and Evercore, we have a simple seven-step playbook to help harness the energy between our ears. We don't want anyone to lose even one megawatt of power because of the distractions of their past, the present, or worries of their future. So, we encourage you to start with this easy path to harness the human factor of energy innovation.

1. **Practice Mental Fitness and Resilience.** Invest in coaching, mental health, and resources for your team and their families.

2. **Know Yourself.** Get clear on your mission, vision, values, and unique ability.

3. **Individual flow.** Work on individual flow first. If you can't get into individual flow, group flow is likely out of reach.

4. **Group Flow.** After you reach individual flow, you, as a visible leader, can push yourself and your teams just beyond comfort—not too easy, not too much—to create group flow.

5. **Lead with Humility and Persistence.** Great teams in flow always have a humble leader holding it together, making the hard decision to keep the right cats in the right seats.

6. **Use Technology as an Amplifier.** AI and chatbots don't replace leaders or people—they empower, supercharge, and scale your impact.

7. **Share Your Model.** The world needs more mental fitness, not just more technology. Don't hoard what works. Lead by example.

The energy revolution is not just about technology, capital, or regulation; it's about the power within each person, team, and leader. Unlock that, and you unlock abundance everywhere. You will come in and out of flow. It's not a constant; however, practice and intention will make it more common.

The Bottom Line:
The most valuable power plant you
own is your brain.
Mental fitness is not just a perk,
it's the foundation for innovation, safety,
and everything you'll build.
Culture is strategy: Know Thyself, Love Thyself
and then lead others by example.
Group flow is possible,
but only with a great leader who has a clear MVV,
mental fitness, and their own
Unique Ability and passions.
If you want to change the energy industry
—or any industry—
start with yourself, your team, and your tribe.

PART IV

THE MANIFESTO

CHAPTER 9

ENERGY ABUNDANCE—THE ENTREPRENEUR'S MANIFESTO

The story of American energy is the story of builders, not bureaucrats; of men and women who didn't wait for permission. Those pioneers acted, took risks, and delivered what others said was impossible.

It's also the story of my father, Leon Wandler, who began with a welding torch in a Wyoming machine shop and built L&H Industrial to become a global force. It's the story of Jerry Schloredt, Senior Chief Petty Officer, who kept microreactors running in Antarctica—not for headlines, but because he accepted the challenge to keep the lights burning and people safe. It's the story of Chuck Fegley, who led the first Seabee nuclear teams to the bottom of the world, showing what real leadership means when you're utterly alone.

And it's the story of the next generation—Evercore Energy and the entrepreneurs, engineers, and customers who refuse to settle for less than Safe, Quality, Competitive energy.

WHAT IS ENERGY ABUNDANCE?

I wrote this book to share the truth. We live in a world of energy abundance, but it's more than just cheap power. Energy abundance is the freedom to build, grow, and serve others without fear of scarcity or sudden disaster. It's the security of knowing that your family, your company, or your community can survive the next storm, drought, or blackout. It's the dignity of real work and the pride of delivering value that lasts. It's a continuation of those of us in the USA being spoiled by the power always being there and affordable.

We don't even think about power. And the only way it will continue is if we keep up with this energy demand and finally increase production after forty years of flatlining. But our abundance can be bigger than our selfish interests. I envision energy innovation that provides this same energy security to our allies in the free world. True energy abundance means the USA and its allies provide all energy, including nuclear microreactors, to the rest of the world through power purchase agreements, aggregating demand and becoming the leading energy supplier for every kind of energy.

THE PLAYBOOK FOR BUILDERS—WHAT SETS US APART

L&H and Evercore are inviting anyone who shares our vision to walk with us into the next generation of energy production. But for some, it will require mindset changes. We've been gearing up for this boom for years, and we've developed a mental roadmap for success:

- We say "no" to magical thinking, and "yes" to results.
- We don't demonize carbon, worship renewables, or bet the farm on government bailouts.

- We hold every solution to the same standard: *Would I bet my family's business on this?*
- We listen to the people who pay the bills—the customers, workers, and families, not just the experts and politicians.
- We honor the wisdom of those who came before— *Leon, Jerry, Chuck, and every machinist, welder, operator, and engineer who kept the faith and endured the iterations when others gave up.*

A NEW FUTURE

Those who open the door to energy innovation will create a new and exciting future. Imagine a world where:

- America leads the world in AI and chip-making because we innovate energy that makes it happen at competitive rates.
- Rural hospitals never close because of unreliable power.
- Power can be dispatched where it's needed on demand to the most remote or recently devastated places on earth.
- Mines and mineral processing locate the minerals, and a company like Evercore efficiently and competitively dispatches safe, high-quality energy to supply their demand.
- Innovation is unleashed, and young leaders can build without running into bureaucratic brick walls.
- Energy is not a source of division, but a platform for unity and an opportunity for the free world.
- Hundreds of excellent companies anchored perpetually in the USA and allied countries own and protect

nuclear microreactors, SMR, and other innovative energy. Global power purchase agreements keep energy revenues and profits coming into the countries that share freedoms and safety as core values. These companies are committed to innovating and serving the free world, ensuring national security, aggregating demand to drive competitive costs, and letting free world allies reap the rewards for supplying the world's energy.

LEADERSHIP LESSONS FOR THE NEXT GENERATION

As we step into the next generation of energy, the Three Drivers—Safe, Quality, and Competitive—are imperative. At the same time, there are six basic principles that leaders need to remember if they want exponential success:

1. **The truth will set you free.**

   Energy is an extremely complicated topic. With an array of data—some excellent and some questionable—as well as more than one valid opinion, it can be confusing. However, if you keep an open mind and always search for the truth, you will be ahead of most people. The sticky part of the energy dilemma is that more than one truth can be plausible at the same time. For instance, it's true that nuclear energy has experienced some high-profile fatal accidents. It's also true that, when you compare those fatalities to those of other energy sources, nuclear energy is one of the safest. If you start lying to yourself, your spreadsheet, or your AI, it will compound and feed those lies back to you at an impressive rate.

2. **Build with Integrity.**

   Your name is your word. Don't cut corners. Do what's right, even if it costs more today. It will save lives, reputations, and even money tomorrow.

3. **Demand Real Value.**

   Don't settle for the cheapest, and don't be the cheapest. Be the best. Focus on uptime, reliability, total cost of ownership, and long-term partnerships with like-minded people you trust. The cheapest perpetually means zero profit. It might be the most crowded business area, but it also has the most problems and suffering for all involved. Some jackwagons will always cut corners and offer lower prices to get the sale. Inevitably, it will have to be done again, but if they didn't have funds to do it right the first time, they won't have money to rework or replace whatever went wrong.

4. **Learn from Failure.**

   Every blackout, broken part, or policy blunder is a chance to get better. Never stop learning and never let pride get in the way of progress.

5. **Serve Others.**

   True abundance is not about having more; it's about making sure others can thrive, too. The goal is to lift as you climb. Love is the most valuable thing we can gather and hold on to. When you gather tribes of like-minded people with shared missions, visions, and values, you can be in flow together and experience pure LOVE—Love for your life, your relationships, and your life's purpose.

6. **Lead with Courage.**

In every crisis, someone must take responsibility. Be the leader who stands up, communicates honestly, and earns trust. Excellent leaders will lead the charge to innovate in energy, and we have room for hundreds, if not thousands. Energy knows no scarcity—just abundance—and, as my friend Steve says, "Target-rich environments in every direction."

VISION IN ACTION

L&H Industrial and Evercore are working closely with the Wyoming Innovative Entrepreneurs (WIE) to stay on top of energy solutions and replace the myths of energy scarcity with the truth of innovative energy. We have a vision to become leaders in energy innovation and abundance, and we're looking for people who share our manifesto goals to join us. Here's a brief look at what our vision looks like put into action as we serve today, tomorrow, and through the next generation.

AT L&H INDUSTRIAL:

We see ourselves as part of the nuclear supply chain, providing parts and services to customers that fit our unique capabilities, passions, and resources. As we do for coal, oil, hydro, and wind, we're rebuilding and refreshing legacy nuclear plants.

Our superpower is short-run, custom, safety-critical parts—the kind needed for prototypes, repairs, and legacy equipment where downtime isn't an option. Think of us as the ultimate makers of space for the largest equipment in the world. This gives us an advantage as we begin replacing or competing with original equipment manufacturers or supply chains that have withered or been sold off.

Continuous innovation drives our long-term partnerships. We don't just supply parts; we help owners perpetually modernize and optimize their assets. From engineering and design through manufacturing and repair to field service—from hard iron to digital—when we take over a machine, we get to know it better than anyone because it's our unique capability and passion. Most competitors don't love the machines. We do.

OPS IQ: Led by the next generation of L&H, digital IoT sensors and real-time monitoring transform machines into intelligent systems. I'm certain my father, Leon Wandler, is smiling to see the innovation, entrepreneurship, and work ethic he started continue generationally through Gage and Al Wandler.

OPS IQ Mentor: L&H is building the ultimate knowledge base by feeding manuals and tribal wisdom into AI, so each generation of operators and maintainers can transfer their expertise to the next, faster and better. This is the biggest innovation since electricity.

THE MICROREACTOR OPPORTUNITY & WIE

I'm honored to lead and serve with the Wyoming Innovative Entrepreneurs to grow and diversify Wyoming's economy. Energy innovation works best when entrepreneurs lead the way. Just look at the Bill Aulet MIT results for proof. You can also learn more about WIE at InnovateWY.com.

Working together, L&H and WIE are building a world-class, resilient supply chain for microreactors in Wyoming and across the USA. We dream of every winning microreactor company being able to call on three vetted Wyoming/USA suppliers for every part, every time, and get actual competition, better quality, and more speed. L&H will shine in prototyping and critical, high-spec parts. As the industry scales—imagine fifty or more reactors placed every year—we'll help build automated branches for high-volume production, matching teams to the work that best fits them. WIE opens the door for even more entrepreneurs, allowing customers to get multiple quotes and best-in-class service. WIE has the potential to offer more options, even for the areas where L&H excels. This means the supply chain grows stronger for everyone.

EVERCORE: OUR VISION FOR MICROREACTORS

To every microreactor company and technology developer:

Evercore's vision centers around truly movable microreactors—reactors that can be deployed where our customers need them, scaled up with more modules as demand grows, and then relocated to new sites when customer needs change. We're fine if a move only happens during a fuel change. That fits the practical reality of nuclear energy.

Fixed-plant SMRs have a strong place in the market, and we support their growth for grid and industrial applications. But Evercore's business model—our focus and the promises we make our customers—involves movable, modular, flexible energy. Our business plan envisions us owning fleets of movable microreactors, matching them to customer needs in mining, mineral processing, heavy industry, and national security to help our clients go where the opportunity is, not where the grid happens to exist.

We want to talk to you if you are building a microreactor that can be moved, scaled, and redeployed, even if it's only at refuel cycles. If you share our mission, vision, and values and want to see your technology deployed in the real world, Evercore wants to partner with you, innovate, and grow with you.

As the hungry AI, chip, and industrial sectors aggregate demand and drive prices down, Evercore will be at the leading edge—matching solutions to real customers, gathering financing, managing assets, and keeping energy competitive for the free world. Most importantly, we will work to keep perpetual ownership and the headquarters of these energy assets based in the USA and allied countries, so that freedom will continue to lead the world's values.

WHAT DRIVES US FORWARD

L&H, Evercore, and WIE are committed to working with customers and vendors who share our mission, vision, and values. We consider energy innovation a calling, not just a business. Flow, fun, and purpose are non-negotiables, so we will repel those businesses that don't share our MVV.

Low-hanging fruit and "soft targets" are everywhere; the next twenty-five years will be the most exciting, innovative era we've ever seen. The combination of innovative energy, AI, IoT, and real-time digital information will make finding the truth easy and create abundance.

We invite everyone who is ready to build, innovate, and lead—microreactor companies, suppliers, entrepreneurs, and customers—to connect with us and join this movement.

FINAL THOUGHT

It's time to make the microreactors hum again. Join me in honoring the legacy of Jerry, Chuck, Leon, and all the pioneers who have always pushed the boundaries of what's possible. Innovation has always been the way of the world. The pace has never stopped accelerating.

Today, the compounding and convergence of innovation in energy, AI, manufacturing, and leadership means progress is moving faster than ever. This isn't a new story. It's nature itself: relentless, unstoppable, and always moving forward.

Some people want to slow it down, regulate, or debate progress into the ground. But real innovation, driven by leaders with courage, integrity, and shared values, is a force of nature. It can't be stopped. It compounds, accelerates, and lifts everyone willing to see and act on the vision.

Let's keep pushing. Let's help each other stay in flow. And let's build a future where resilience, abundance, and possibility are not the exception but the standard.

**The world needs more leaders,
more doers and more dreamers
who won't settle for "good enough."
Let's get to work—together.**

APPENDIX

A Short Glossary

Though these are not terms we used in the book, they are acronyms and terms that will be helpful as you explore truth and learn about innovative energy.

FEA (Finite Element Analysis): Computer modeling that shows where a part will bend, break, or fatigue.

WPS (Welding Procedure Specification): The official "recipe" for a weld — materials, filler, heat, speed, position. Must be followed every time.

PQR (Procedure Qualification Record): Proof that a welding procedure has been tested and passed destructive and nondestructive tests.

WPQ (Welder Performance Qualification): Certification showing that a welder can perform a specific procedure correctly.

NDE (Nondestructive Examination): Testing methods (ultrasonic, radiographic, magnetic particle, liquid penetrant, visual) that check quality without damaging the part.

MTR (Mill Test Report): Certificate from a mill proving the chemical and mechanical properties of a piece of metal.

Heat Number: Unique ID stamped on metal to trace it back to the exact batch it came from.

Traveler: The paperwork that follows a part through every stage of manufacturing, recording inspections, tests, and approvals.

CFSI (Counterfeit, Fraudulent, Suspect Items): Parts that aren't what they claim to be. Preventing these is a cornerstone of nuclear-grade manufacturing.

FAT/SAT (Factory Acceptance Test / Site Acceptance Test): Full testing of equipment before shipping (FAT) and again after installation in the field (SAT).

CAPA (Corrective and Preventive Action): A structured process to fix problems and prevent them from happening again.

COMMEMORATION

In January 2010, the Naval Nuclear Power Unit Group Team dedicated a plaque in honor of the men who served McMurdo Station from 1961 through 1972. Phil Smith served in the Antarctic and spearheaded the endeavor after realizing most who had served "felt as though they had been erased from the history book and forgotten."[6] Chuck Fegley supplied technical and historical information, and the team, which included Phil and Chuck, as well as Bob Garland, Leonard McGregor, Rex Hoover, Guy Guthridge, Arden Bement, Karl Erb, Dave Breshnan, Brian Stone, and George Blaisdell, raised funds and took care of all the logistics to ensure their work, in the harshest conditions, which resulted in never having a release of radiation in excess of safety levels, would be commemorated forever. The Secretariat of the Antarctic Treaty described the work in this way:

"The PM-3A nuclear reactor was the first, and only, experiment to power an Antarctic station with a nuclear reactor. The motivation was to reduce the reliance on fuel oil at McMurdo Station. The PM-3A arrived at McMurdo Station on December 12, 1961, and began producing electricity for the station on July 10, 1962. The 1.8-megawatt reactor was decommissioned when continued operation would no longer be cost-effective. The disassembly and removal of the station and most of the associated buildings continued until 1979, when a radiological survey, as well as a subsequent review commissioned by the US Navy, determined the radiation levels at the site were similar to background radiation levels, and there was minimal risk from radiation exposure. The US Department of Energy then released the site for unrestricted use. The last remaining buildings were removed during the 2009-2010 austral summer."[7]

Learn More: Microreactors, Past and Present

- Microreactors: Looking to the Past to Power the Future (INL Video)
- PM-3A Microreactor History (Wikipedia)
- Schloredt Nunatak, Antarctica (Gazetteer)
- Jerry Schloredt Obituary & Legacy (Seabee.org)

ENDNOTES

1   *Insights Offered by Mitrade.* "AI data centers may consume more power than whole cities." November 23, 2024. https://www.mitrade.com/insights/news/live-news/article-3-485463-20241123.

2   Cohn, Lisa. *Microgrid Knowledge.* "44% of Businesses Considering Microgrids: Survey." July 17, 2020. https://www.microgridknowledge.com/distributed-energy/article/11428781/44-of-businesses-considering-microgrids-survey.

3   Ritchie, Hannah. *Our World Data.* "What are the safest and cleanest sources of energy?" February 10, 2020. https://ourworldindata.org/safest-sources-of-energy.

4   StartUpArchive. *X.* "Elon Musk explains his 5-step algorithm for running companies." December 27, 2024. https://x.com/StartupArchive_/status/1872625977672831146.

5   Ritchie, Hannah. *Our World Data.* "What are the safest and cleanest sources of energy?" February 10, 2020. https://ourworldindata.org/safest-sources-of-energy.it ,

6   Rejcek, Peter. *Antarctic Sun.* "Powerful Reminder." June 25, 2010.

7   *Secretariat of the Antarctic Treaty.* "HSM 85: Plaque Commemorating the PM-3A Nuclear Power Plant at McMurdo Station." Accessed October 16, 2025. https://www.ats.aq/devph/en/apa-database/160.

ACKNOWLEDGMENTS

No one builds alone. My work, this book, and every part of my life stand on the shoulders of giants—mentors, friends, partners, and above all, family.

TO MY FAMILY

To my wife, **Jerri Wandler**—the Love of my life and the greatest partner I could ever ask for. Your encouragement, patience, and relentless support make this (and everything else I dream up) possible. I love you more than you'll ever know.

To my father, **Leon Wandler**—whose entrepreneurial grit and belief in building things that last have shaped everything I do. Your hands built L&H; your values built me. Every chapter in this book, and every leap we take, traces back to you.

To my sons, **Gage Wandler and Al Wandler, and nephews Jason Percifield and Dustin Roush**—the next generation of builders, innovators, and leaders. Watching you carry forward the spirit of L&H and push it into new frontiers is the greatest legacy I could hope for. Your work on OPS IQ, OPS Mentor, digital transformation, and the next wave of innovation inspires me.

And to **all the L&H tribe**—past, present, and future— thank you for making it possible for me to chase big ideas, for challenging me, and for reminding me that love for our tribes of people and for the lives we create together is always the real "why."

TO MARCIO PAES BARRETO

To Marcio—my friend, early Evercore partner, and original co-writer of this book; a relentless thinker, and a true leader in energy, history, policy, and statistics.

Marcio joined Evercore in its infancy and played a critical role in shaping both the company's foundation and this manuscript. Many of the historical insights and policy perspectives in these pages came from his deep research and hard work. Our partnership has been pivotal to Evercore's beginnings. Marcio's work ethic and passion are truly exceptional—he worked day and night, many weekends, and poured everything he had into building something remarkable from scratch. His commitment, intellectual honesty, and persistence set a standard for what partnership can mean in a startup.

As 2025 comes to a close, our paths are beginning to diverge—because microreactor technology, competitive pricing, and delivery likely won't be ready until around 2030. It wouldn't make sense for Marcio to wait years for the microreactor market to mature when his ideas for grid innovation are ready to blossom in his new startup and future projects. Meanwhile, my vision for Evercore remains clearer than ever: to understand, match, and manage the perfect energy solutions for my customers in mining, mineral processing, and beyond—and ultimately to own nuclear microreactors, partnering with the best operators and brands, and patiently waiting for all these fantastic microreactor entrepreneurs to catch up.

When Marcio and I set out to write this book together, we quickly realized we were coming at it from two very different angles. My style is quick-start, entrepreneurial, and focused on momentum—never worrying if every detail is perfect, as long as the mission, vision, and values are moving forward and what I'm saying is true. Marcio is a deep researcher, a fact-checker, and a professor-level intellect who values precision and integrity in every word. Steve Aumeier, who has also shaped so much of my understanding, shares that same scientific honesty and depth.

As Marcio decided to step out of the day-to-day at Evercore Energy, it hit me that I would be the one speaking about this book, and it needed to be in my words, from my head, heart, and gut, so I could deliver it in a flow. I want readers to know that any "fast and loose" commentary, entrepreneurial leaps, or broad statistics on these pages are mine alone. Neither Marcio nor Steve is responsible for any statements that run a little too "fast and loose" or simplify complexity for the sake of momentum. They are both extraordinary researchers and far more knowledgeable than I will ever be on these topics. When things get serious, I call them.

My Unique Ability® is to learn from people like Marcio and Steve and put those lessons into action—to lead, build, iterate, and drive real change for my customers and teams. I'm not afraid to be wrong, and I welcome being challenged and corrected. That's how I learn. I'll leave the perfect facts and the pure teaching to the PhDs and professors. My style is for entrepreneurs and my tribes—people who want to move, build, and make things better rather than wait for the "perfect" solution to arise.

Marcio's contribution to this book and to Evercore will have a lasting impact on the next wave of energy innovation, and we intend to stay connected and find new ways to innovate in energy together.

TO DR. STEVE AUMEIER

Steve's lifelong work in nuclear engineering, his technical expertise, and above all, his commitment to teaching and mentorship have deeply influenced both Marcio and me on this journey. If you see honest science, engineering rigor, and technical grounding in these pages, much of it is thanks to Steve's guidance and generosity.

Steve is not just a world-class scientist and builder—he's a friend and mentor who always made time for my questions, no matter how "out of left field," they were. He gave me the confidence to keep learning, to push past my limits, and to see possibilities in nuclear energy innovation that most people never consider. He leads with humility, relentless curiosity, and a deep belief that the next generation of energy leaders is already out there—if we empower them to build.

Professionally, Steve Aumeier holds a Ph.D. in Nuclear Engineering from the University of Michigan, an MBA from the University of Chicago, and serves as Senior Advisor for Strategic Programs at Idaho National Laboratory (INL)—one of the world's premier nuclear research institutions. He previously led INL's entire Energy and Environment Science & Technology Directorate, guiding over 300 scientists and engineers working at the front lines of clean energy, advanced nuclear, national security, and sustainability. His career also includes significant roles at Argonne National Laboratory and deep partnerships across government, academia, and industry.

When I say, "I stand on the shoulders of giants," I'm thinking of people like my father and Steve Aumeier. Steve is a teacher's teacher. Any technical accuracy, safety obsession, or intellectual honesty you see here—credit Steve. Any "fast and loose" entrepreneurial leaps, oversimplifications, or statistics stretched for effect—those are mine.

Steve's contribution to me, Marcio, and the State of Wyoming is lasting. His influence goes far beyond this book, and his encouragement has made all the difference in my journey as a business leader and a student of energy.

ABOUT THE AUTHOR

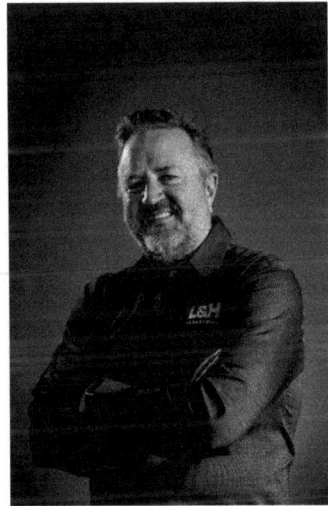

Mike Wandler prioritizes love, freedom, mental and physical fitness, energy, and flow. He serves as President of L&H Industrial, Inc., a third-generation family enterprise committed to its mission of moving industries forward by designing, manufacturing, and maintaining the biggest machines on earth. Under Mike's leadership, L&H delivers engineering, manufacturing, repair, and field services to Mining, Oil & Gas, Railroad, Wind Energy, Government, and other heavy industrial sectors—guided by the company's core values: Do the Right Thing, Lead by Example, Make an Impact, and Love People.

Mike is also the CEO of Evercore Energy, an energy-innovation company advancing the deployment of microreactors and pioneering reliable, sustainable, data-driven energy systems for heavy industry. His work centers on transforming how the world powers essential industrial operations.

As Founder of Wyoming's Innovative Entrepreneurs, Mike champions statewide economic growth by helping entrepreneurs think bigger, access better resources, and build resilient companies.

Mike's unique ability is his relentless pursuit of knowledge, technology, and simple truths; his talent for expanding visions; and his instinct for pioneering practical, innovative solutions. Whether in business, mechanics, or leadership, he builds cultures grounded in honesty, integrity, respect, safety, and quality. He believes that transformative progress comes from living in flow, surrounding himself with exceptional people, and collaborating with those who share his values. He continually sharpens his craft through programs including Harvard OPM38, MIT's Entrepreneurship Development Program, Strategic Coach FreeZone, and Abundance 360 (Patron).

Most importantly, Mike's journey has never been a solo pursuit. His wife, Jerri Wandler, has been his unwavering partner—his grounding force, his inspiration, and the person who has made every breakthrough in his life possible. Her love, strength, and devotion are woven into every chapter of his story. Together, they have raised seven children and celebrate the joy of fourteen grandchildren, each a reminder of what legacy truly means.

Above all achievements—beyond business, machinery, or energy innovation—Mike's deepest purpose is rooted in love, freedom, and flow. These principles guide his decisions, shape his leadership, and anchor the life he and Jerri have built together.

CONNECT WITH MIKE

@mike-wandler-lh-industrial

Follow him on LinkedIn today.

L&H
INDUSTRIAL

L&H Industrial is a leader in technology innovations, custom manufacturing, and comprehensive services for heavy industrial machinery used in heavy industries across the globe.

LNH.net

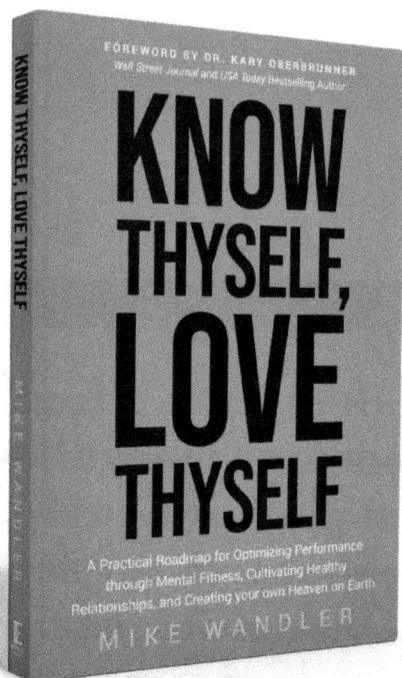